Taniguchi Symposia on Brain Sciences No.8

ENDOGENOUS SLEEP SUBSTANCES AND SLEEP REGULATION

Taniguchi Symposia on Brain Sciences No. 8
PROGRAM COMMITTEE

Shojiro Inoué (Chairman), Hachiro Nakagawa,
Yasuro Takahashi, Shizuo Torii, Koji Uchizono

The Taniguchi Foundation, Division of Brain Sciences
ORGANIZING COMMITTEE

Taniguchi Symposia on Brain Sciences No.8

ENDOGENOUS SLEEP SUBSTANCES AND SLEEP REGULATION

Edited by
Shojiro Inoué
and Alexander A. Borbély

CRC Press
Taylor & Francis Group
Boca Raton London New York

CRC Press is an imprint of the
Taylor & Francis Group, an **informa** business

Published jointly by
JAPAN SCIENTIFIC SOCIETIES PRESS Tokyo
ISBN 4-7622-9469-1
        and
VNU SCIENCE PRESS BV Utrecht, The Netherlands
ISBN 90-6764-058-1

Distributed in all areas outside Japan and Asia between Pakistan and Korea by VNU Science Press BV Utrecht, The Netherlands

PREFACE

What is sleep? What is an endogenous sleep substance? In spite of the efforts dating back to ancient Greece and China to answer these questions, they remain still open to day. Neither the term 'sleep' nor the concept of a 'sleep substance' has been unambiguously defined. Both lie still in almost unexplored regions of present-day science. Scientists should endeavour to open the door of this unknown world by applying new concepts and modern techniques.

Based on objective EEG monitoring, a pioneer search for an endogenous sleep factor was initiated by Marcel Monnier and his associates in Switzerland in the early 1960s. They were followed by John R. Pappenheimer and his collaborators in United States and by Koji Uchizono and his research team in Japan. Each of these ambitious projects was rewarded by the isolation and identification of an active substance. In the wake of these studies, the classical concept of humoral sleep regulation has witnessed a revival in the recent years. The dramatic progress in many areas of the neurosciences contributed to the growing interest in sleep mechanisms and sleep substances. Consequently, the time seemed ripe to take stock of the 'state of art.'

An international meeting was planned to discuss the major recent developments and to examine their conceptual background.

It was extremely fortunate that the Taniguchi Foundation accepted to sponsor the symposium. This foundation, created by Mr. Toyosaburo Taniguchi in 1929, supports every year a number of small-scale international meetings in order to further the progress of basic sciences and to promote mutual friendship among promising young scientists. The Organizing Committee of the Division of Brain Sciences, headed by Dr. Osamu Hayaishi, decided to adopt a meeting entitled "Humoral Control of Sleep and Its Evolution" at the proposition of Shojiro Inoué as the 8th Taniguchi International Symposium. The Program Committee, chaired by Inoué in cooperation with Hachiro Nakagawa, Yasuro Takahashi, Shizuo Torii and Koji Uchizono, carefully selected the topics and speakers, and organized the symposium in Kyoto/Katata, Japan, on October 21–25, 1984.

This book is composed of 21 review articles based on the presentations at the symposium, in which the authors review and interpret their recent experimental results on sleep substances. Some of the papers focus on regulatory mechanisms and the functions of sleep with special reference to the somnogenic activities of humoral factors. Evolutionary and adaptive aspects, and the circadian rest-activity rhythm constitute further important topics. The variety of different substance candidates and the multidisciplinary approaches are integral features of this rapidly developing research area. This is the first book which is entirely devoted to this topic and attempts to provide a representative overview of important trends in the field. The editors hope that this volume will promote interests in these new developments, and will ultimately contribute to the understanding of sleep and its humoral regulation.

January 1985

S. Inoué
A.A. Borbély

CONTENTS

GENERAL ASPECTS

1

SLEEP SUBSTANCES: THEIR ROLES AND EVOLUTION

SHOJIRO INOUÉ

Division of Biocybernetics, Institute for Medical and Dental Engineering, Tokyo Medical and Dental University, Tokyo 101, Japan

I. HISTORICAL BACKGROUND

As early as November 1906, the French psychophysiologists René Legendre and Henri Piéron initiated sleep deprivation in dogs in order to obtain an endogenously occurring sleep-inducing substance. In July 1907, a Japanese physicochemist, Kuniomi Ishimori, undertook almost the same experiments in dogs. The former published a series of short reports in French during 1907–1912 and a final full paper in 1913 (*24*), while the latter published a 29-page full paper in Japanese in 1909 (*19*). Both of the pioneer researchers independently reached the conclusion that a long-lasting state of wakefulness eventually causes the accumulation of a toxic substance in the brain which may trigger sleep.

Subsequent attempts to search for sleep substances were unsuccessful until the early 1960's, as critically reviewed by Kleitman (*20*). However, a modern study on sleep substances had already been initiated by Kornmüller *et al.* (*22*). They transfused the blood from one cat to another by connecting the carotid arteries. Both cats had been chronically implanted with stimulating and recording electrodes in the brain.

3

The state of sleep/wakefulness was monitored by electroencephalogram (EEG). The authors noticed that the initiation of sleep induced by electric stimuli to the thalamus of one cat soon caused sleep in the other animal. From this result the German neurophysiologists concluded that a blood-borne sleep hormone might be involved in the induction of sleep. After confirming these observations in the rabbit, Monnier *et al.*

TABLE I

List of Putative Sleep Substances

Name of substance	Chemical class	Site of presence	Animal species tested	Reference[a]
Adenosine	Nucleoside	Not specified	Rat	*34*
Arginine vasotocin	Nonapeptide	Pineal	Cat, human, rat	*32*
Benzodiazepine receptor agonists	Unknown	Brain	Rat	*38*
DSIP	Nonapeptide	Blood, brain, pineal, *etc.*	Cat, human, rabbit, *etc.*	*7, 37,* 12–14
Factor S[b]	Muramyl tetra-peptide	Brain, CSF, urine	Rabbit, rat	*31,* 15
Gamma-brom[c]	Organic bromine compound	CSF	Cat	*41,* 20
Insulin[b]	Protein	Pancreas	Rat	*5*
Interleukin-1	Unknown	Blood, brain	Rabbit, rat	*23,* 15
Melatonin	Indolamine	Pineal	Cat, human, rat	*8*
Peptide-like factor[b]	Peptide	Urine	Rat	*44*
Piperidine	Amine	Brain	Cat, human	*25*
PGD$_2$	Cyclic fatty acid	Brain, pineal, *etc.*	Rat	*42, 43,* 16
PS factor[c]	Unknown	CSF	Rat	*1,* 19
REM-protein[c]	Macromolecular protein	CSF, blood	Cat, rat	*39*
REM sleep factor[c]	Unknown	CSF	Cat	*36*
Sleep hormone	Unknown	Blood	Cat	*22*
Sleep-inducing substance	Unknown	Urine	——	4
SPS	Four components	Brain	Crayfish, mouse, rat	*12, 14, 27,* 1, 17
Tryptophan	Amino acid	Not specified	Human, rat, *etc.*	*33,* 14
Uridine	Nucleoside	Brain	Beetle, fish, mouse, rat	*10, 13, 21,* 1, 17
Vasoactive intestinal polypeptide[c]	Polypeptide	Brain, gut	Rat	*35*

[a] Underlined numerals indicate the chapter number in this book.

[b] Effective in increasing SWS only.

[c] Effective in increasing PS only.

CSF, Cerebrospinal fluid.

(*26*) undertook an extensive search for the sleep-inducing substance which finally yielded the delta-sleep-inducing peptide (DSIP).

Since the isolation and identification of DSIP by Schoenenberger *et al.* (*37*) in 1977, the concept of a humoral control of sleep has been vigorously revived and its validity rapidly established. Sleep mechanisms, still a mystery of modern science, could now been discussed on the premise of specific endogenous sleep substances. Consequently, a number of endogenous factors were nominated as putative sleep substances (Table I), although their particular roles in the regulatory mechanism of sleep remain unknown. Reviews of the recent progress in the research of sleep substances have been published by Inoué *et al.* (*18*), Borbély (*2*), Drucker-Colin and Valverde-R. (*6*), Inoué (*11*), Ursin (*44*), Uchizono (*40*), and Graf and Kastin (*7*).

II. COMPARISON OF SOMNOGENIC PROPERTY IN UNRESTRAINED RATS

1. A Technique for the Standardization in Sleep Assay

One of the technical problems in the search for a sleep substance consists in the reliable detection of a sleep-enhancing effect and its evaluation in quantitative terms. Most workers have adopted a short-term bioassay. However, sleep is largely modified by the circadian clock mechanism (*3*). Hence, the day-to-day dynamics in sleep parameters should be also taken into account in sleep bioassay.

We have developed a long-term bioassay technique (*9*) for the quantitative evaluation of the somnogenic property of sleep-promoting substance (SPS) which was extracted from the brainstem of sleep-deprived rats (*27*). A test material can be steadily infused into the third ventricle of otherwise saline-infused freely moving rats. A 72-hr to 96-hr continuous polysomnogram is analyzed with respect to the modulatory effect on the circadian sleep pattern. Since the circadian sleep-waking cycle of the rat is considerably stable even under steady saline infusion (*12*), a reliable and reproducible result can be expected by this routine technique.

2. Differential Roles of Sleep Substances

Five candidates were examined by the above technique (*15*, *16*). They

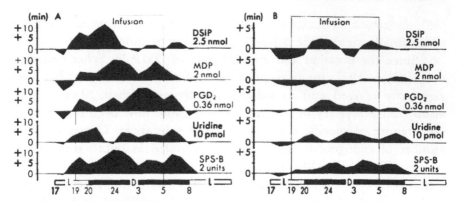

Fig. 1. Comparison of the somnogenic property of five substances nocturnally infused into the third ventricle of freely moving rats. Curves show hourly increments in SWS (A) and PS (B) expressed as difference from the baseline value obtained on the previous day for saline infusion. The total dosage in a 10-hr infusion period is indicated under the name of each substance. L and D represent the environmental light and the dark period, respectively.

were DSIP, muramyl dipeptide (MDP), prostaglandin D_2 (PGD$_2$), uridine, and SPS-B. MDP is a synthetic substance closely resembling factor S which was first detected in the cerebrospinal fluid of sleep-deprived goat (*31*) and recently identified as a muramyl tetrapeptide (see Chapter 15). PGD$_2$ is a natural constituent of the brain and proved to be a potent sleep inducer (*42*, see also Chapter 16). Uridine is one of the active components of SPS (*21*). SPS-B is an unidentified active component of SPS (*12, 14*).

A 10-hr nocturnal intracerebroventricular infusion of an optimal dose of each substance revealed compound-specific somnogenic characteristics in both slow wave sleep (SWS) and paradoxical sleep (PS) (Fig. 1). DSIP was immediately effective and very potent in inducing excessive sleep but the effect was short-lasting. Even during the course of infusion, DSIP ceased to elicit the effect. In this respect, DSIP might be regarded as a primary sleep-inducer or a trigger substance of sleep. MDP was characterized by its slow SWS-promoting effect. The maximal effect was observed in the middle of the infusion period, when a marked elevation of brain temperature was observed (*16*). The sleep-enhancing effect of PGD$_2$ was apparent at the beginning of the infusion period, but was even more prominent during the later period. Thus, PGD$_2$ may primarily induce sleep and secondarily activate sleep-main-

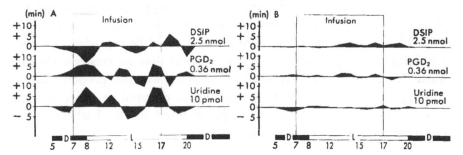

Fig. 2. Comparison of the somnogenic property of three substances diurnally infused into the third ventricle of freely moving rats. For explanations, see Fig. 1.

taining mechanisms. Uridine and SPS-B were characterized by their steady, long-lasting sleep-promoting effects. Both SWS and PS were affected. These SPS components seem to play a more or less similar basic role in triggering and maintaining sleep (see also Chapter 17).

3. Circadian Variations in Effectiveness

Since circadian variations are well known in the effectiveness of exoge-nously administered drugs, the question arises as to whether a diurnal and nocturnal administration of sleep substances induce similar sleep-promoting effects in rats. Hence the 10-hr infusion period of various substances was shifted by 12 hr. In dramatic contrast to the nocturnal administration, the diurnal administration resulted for all the substances in little sleep-modulating effect (15) (Fig. 2).

A possible cause for this marked difference in effectiveness may derive from circadian variations in the requirement for sleep. During the light period, the physiological demand for sleep can be expected to be high enough that the neurohumoral system serving sleep regulation is maximally operative and hence all the time available for rest is oc-cupied by sleep. Accordingly, the exogenously supplied sleep substance does not compete with the abundantly liberated endogenous factors and therefore no extra sleep is detectable. In contrast, during the dark period, sleep is normally suppressed and replaced by wakefulness. The time of wakefulness can be easily curtailed by administering an exogenous factor which evokes sleep.

At present, there is no evidence for a negative feedback mechanism in the humoral control of sleep. However, if it is assumed that sleep in rats is saturated in the light period, no excess sleep is physiologically required. Therefore, if excessive sleep were indeed induced in the light period by an exogenous sleep substance, it could be regarded as a non-physiological sleep. It should be emphasized that an exogenously supplied "endogenous sleep substance" should have the property of not causing extra sleep when the physiological demand for sleep is saturated. This criterion should be supplemented to the Borbély-Tobler requirements for a sleep substance (*4*).

III. EVOLUTIONARY ASPECTS OF SLEEP REGULATION

1.   Phylogenetic Approach to the Rest-Activity in Lower Animals
A)   Rest-activity rhythm versus sleep-waking cycle
The putative sleep substances are found in various organs or tissues of mammals (Table I). Although they do not seem to exhibit species specificity, at least among animals tested, their effects on lower vertebrates and invertebrates are virtually unknown. Since there is much debate about the existence of sleep in non-homoiothermic animals, it is of considerable interest from a comparative physiological viewpoint to examine whether or not sleep substances detected in mammals can exert any activity-modifying effect in lower animals. Accordingly, attempts were made to investigate the effects of sleep substances in fish and insects. The preliminary experimental data which will be presented suggests that the rest-activity regulating mechanism in lower animals may be similar to the mammalian sleep-waking mechanism.
B)   Swimming activity in fish after rest deprivation and administration of sleep substances
Two species of cyprinids, the crucian carp and the topmouth gudgeon, with day-active swimming rhythms were deprived of rest by continuous water streaming which was produced by the rotation of a magnetic stirrer (*13*). Swimming activity was detected by optical switches placed around the vessel. A 24-hr rest deprivation caused a marked reduction in swimming activity in most animals. This may be comparable to the change occurring after sleep deprivation in mammals (see Chapter 5). Consequently, the question arises whether a similar reduction in swimming activity is also induced by the administration of sleep substances.

Fish were exposed for 24 hr to DSIP, a phosphorylated analogue of DSIP, uracil or uridine. A preliminary study (*13*) indicated that all substances were effective either in lowering or activating swimming activity. A dose-response relation was evident for uridine: the maximal reduction in swimming activity was induced by a 100 μM dose, while lower (0.02–10 μM) or higher doses (1 nM) were less effective or even caused an increase in activity. The activity reducing action was comparable to the quiescence induced by 24-hr rest deprivation. The results suggest that the rest-activity regulating mechanism in fish might be basically similar to the mammalian sleep-waking regulating mechanism.

C) Effect of uridine on the locomotor activity in rhinoceros beetles
Giant Japanese rhinoceros beetles with a prominent night-active locomotor rhythmicity were intraabdominally injected with uridine. Lower doses (1–10 pmol) caused a reduction in nocturnal locomotor activity, while larger doses (0.1–10 nmol) induced hyperactivity. Thus uridine can modify the circadian rest-activity rhythm in insects.

2. Evolution of Sleep Substances

The varying features of sleep substances do not allow us to obtain a unitary view of their mode of action. The differential sleep-modulating properties of the various candidate substances may indicate that sleep is regulated by a number of humoral factors. However, the question arises as to why so many sleep-related substances exist, how they participate in the regulation of sleep, and what the specific role of each of them is. The circadian variations in the somnogenic property lead to the further question about the relationship between the sleep-regulating mechanism and the circadian clock mechanism which is so common in biological organisms. Moreover, the problem must be addressed why sleep substances modify rest-activity rhythms in lower animals. Our present knowledge is insufficient to reach a definite conclusion. However, taking into consideration the evolutionary aspects of the neuroendocrine control system in higher animals may provide a possible explanation (*11*).

The higher vertebrates are evolved, the more humoral messengers are involved in the elaborate control of the body. Such substances are mediated by the circulatory system of the body fluid which plays a complementary role to neural regulation (Fig. 3A). It may be postulated that sleep-specific messengers are also in similar ways endogenously

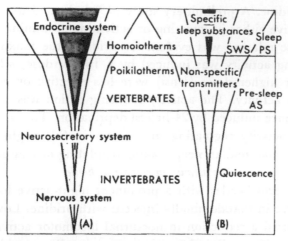

Fig. 3. Schematic representation showing the evolution of the neuroendocrine system and the sleep-regulating system. The higher stage of evolution requires the involvement of humoral control mechanisms. See text.

produced and distributed to the entire central nervous system *via* the cerebrospinal fluid. Such a neurohumoral regulatory system of sleep may have evolved from the basic circadian rest-activity mechanism in parallel with the development of sleep, which represents a recently acquired function in higher animals (Fig. 3B). This speculation is supported by the fact that benzodiazepine receptors are detected in animals higher than bony fish but not lower than jawless fish (*29*).

Sleep control mechanisms are still developing. A number of different substances are utilized to regulate sleep. Even bacterial muramyl peptides are adopted as a "sleep vitamin" (Chapter 15). Therefore, no single sleep-specific humoral mechanism has been established. All putative sleep substances seem to be inadequate for regulating sleep alone. They act by modifying the induction, maintenance and/or development of sleep. Some of them affect both SWS and PS, while others modify either SWS or PS. A further evolutionary step seems to be required to achieve a true sleep-regulating hormone which is no longer influenced by circadian mechanisms (*3*), body temperature regulation (*30*), and the circulating nutrient level (*28*).

SUMMARY

The isolation and identification of DSIP has revived the concept of the humoral control of sleep. To date a number of sleep substance candidates have been nominated. Using the long-term intracerebroventricular infusion technique compound-specific somnogenic effects were observed for putative sleep substances such as DSIP, MDP, PGD_2, SPS-B, and uridine, after nocturnal administration to freely moving rats. In contrast, diurnal administration of the same substances failed to cause excess sleep. The sleep substances also exerted a modulatory effect on the rest-activity behavior of two species of fish, the crucian carp and the topmouth gudgeon, and a species of insect, the rhinoceros beetle. The differential time-dependent sleep-modulating property and the rest-activity regulating property of sleep substances indicate that the neurohumoral regulatory system of sleep may have evolved from the circadian rest-activity mechanism in parallel with the development of sleep, and that sleep is regulated by a number of humoral factors each of which plays a specific role in the dynamic process of sleep.

REFERENCES

1 Adrien, J. and Dugovic, C. (1984). *Eur. J. Pharmacol.* **100**, 223–226.
2 Borbély, A.A. (1982). *Trends Pharmacol. Sci.* **3**, 350.
3 Borbély, A.A. (1982). *Human Neurobiol.* **1**, 195–204.
4 Borbély, A.A. and Tobler, I. (1980). *Trends Pharmacol. Sci.* **1**, 356–358.
5 Danguir, J. and Nicolaidis, S. (1984). *Brain Res.* **306**, 97–103.
6 Drucker-Colin, R. and Valverde-R., C. (1982). In *Sleep—Clinical and Experimental Aspects*, ed. Ganten, D. and Pfaff, D., pp. 37–81. Berlin: Springer-Verlag.
7 Graf, M.V. and Kastin, A.J. (1984) *Neurosci. Biobehav. Rev.* **8**, 83–93.
8 Holmes, S.W. and Sugden, D. (1982). *Br. J. Pharmacol.* **76**, 95–101.
9 Honda, K. and Inoué, S. (1978). *Rep. Inst. Med. Dent. Eng.* **12**, 81–85.
10 Honda, K., Komoda, Y., Nishida, S., Nagasaki, H., Higashi, A., Uchizono, K., and Inoué, S. (1984). *Neurosci. Res.* **1**, 243–252.
11 Inoué, S. (1983). *Seikagaku* **55**, 445–460 (in Japanese).
12 Inoué, S., Honda, K., and Komoda, Y. (1983). In *Sleep 1982*, ed. Koella, W.P., pp. 112–114. Basel: Karger.
13 Inoué, S., Honda, K., and Komoda, Y. (1984). *Abstr. 10th Int. Congr. Biometeorol.*, p. 156.
14 Inoué, S., Honda, K., and Komoda, Y. (1985). In *Sleep: Neurotransmitters and Neuromodulators*, ed. Wauquier, A., Gaillard, J.M., Monti, J.M. and Radulovacki, M., pp. 305–318. New York: Raven Press.

15 Inoué, S., Honda, K., Komoda, Y., Uchizono, K., Ueno, R., and Hayaishi, O. (1984). *Neurosci. Lett.* **49**, 207–211.

16 Inoué, S., Honda, K., Komoda, Y., Uchizono, K., Ueno, R., and Hayaishi, O. (1984). *Proc. Natl. Acad. Sci. U.S.* **81**, 6240–6244.

17 Inoué, S., Honda, K., Nishida, S., and Komoda, Y. (1983). *Sleep Res.* **12**, 81.

18 Inoué, S., Uchizono, K., and Nagasaki, H. (1982). *Trends Neurosci.* **5**, 218–220.

19 Ishimori, K. (1909). *Tokyo Igakkai Zasshi* **23**, 429–457 (in Japanese).

20 Kleitman, N. (1963). *Sleep and Wakefulness*, 2nd ed. Chicago: Univ. Chicago Press.

21 Komoda, Y., Ishikawa, M., Nagasaki, H., Iriki, M., Honda, K., Inoué, S., Higashi, A., and Uchizono, K. (1983). *Biomed. Res.* **4** (Suppl.), 223–227.

22 Kornmüller, A.E., Lux, H.D., Winkel, K., and Klee, M. (1961). *Naturwissenschaften* **48**, 503–505.

23 Krueger, J.M., Walter, J., Dinarello, C.A., Wolff, S.M., and Chedid, L. (1984). *Am. J. Physiol.* **246**, R994–R999.

24 Legendre, R. and Piéron, H. (1913). *Z. Allgem. Physiol.* **14**, 235–262.

25 Miyata, T., Kamata, K., Nishikibe, M., Kasé, Y., Takahama, K., and Okano, Y. (1974). *Life Sci.* **15**, 1135–1152.

26 Monnier, M., Koller, T., and Graver, S. (1963). *Exp. Neurol.* **8**, 264–277.

27 Nagasaki, H., Iriki, M., Inoué, S., and Uchizono, K. (1974). *Proc. Japan Acad.* **50**, 241–246.

28 Nicolaidis, S. and Danquir, J. (1984). *Exp. Brain Res.* (Suppl.) **8**, 173–187.

29 Nielsen, M., Braestrup, C., and Squires, R.F. (1978). *Brain Res.* **141**, 342–346.

30 Obal, F. Jr. (1984). *Exp. Brain Res.* (Suppl.) **8**, 157–172.

31 Pappenheimer, J.R., Miller, T.B., and Goodrich, C.A. (1967). *Proc. Natl. Acad. Sci. U.S.* **58**, 513–517.

32 Pavel, S., Psatta, D., and Goldstein, R. (1977). *Brain Res. Bull.* **2**, 251–254.

33 Radulovacki, M. (1984). *Neurosci. Biobehav. Rev.* **6**, 421–427.

34 Radulovacki, M., Virus, R.M., Djuricic-Nedelson, M., and Green, R.D. (1984). *J. Pharmacol. Exp. Therap.* **228**, 268–228.

35 Riou, F., Cespuglio, R., and Jouvet, M. (1982). *Neuropeptides* **2**, 265–277.

36 Sallanon, M., Buda, C., Janin, M., and Jouvet, M. (1982). *Brain Res.* **251**, 137–147.

37 Schoenenberger, G.A., Maier, P.F., Tobler, H.J., and Monnier, M. (1977). *Pflügers Arch.* **369**, 99–109.

38 Skolnick, P., Mendelson, W.B., and Paul, S.M. (1981). In *Psychopharmacology of Sleep*, ed. Wheatley, D., pp. 117–134. New York: Raven Press.

39 Spanis, C.T., Gutierrez, M.C., and Drucker-Colin, R.R. (1976). *Pharmacol. Biochem. Behav.* **5**, 165–173.

40 Uchizono, K. (1984). *Nihon Rinsho* **42**, 989–1006 (in Japanese).

41 Torii, S., Mitsumori, K., Inubushi, S., and Yanagisawa, I. (1973). *Psychopharmacologia* **29**, 65–75.

42 Ueno, R., Honda, K., Inoué, S., and Hayaishi, O. (1983). *Proc. Natl. Acad. Sci. U.S.* **80**, 1735–1737.

43 Ueno, R., Ishikawa, Y., Nakayama, T., and Hayaishi, O. (1982). *Biochem. Biophys. Res. Commun.* **109**, 576–582.

44 Ursin, R. (1984). *Exp. Brain Res.* (Suppl.) **8**, 118–132.

2

APPROACHES TO SLEEP REGULATION

ALEXANDER A. BORBÉLY

Institute of Pharmacology, University of Zürich, CH-8006 Zürich, Switzerland

Scientists adopt divergent attitudes towards sleep research. Some of them, focussing their attention on the unresolved problems regarding functions and mechanisms of sleep, are critical towards research efforts in this field. Others are impressed by the recent new insights and anticipate a rapid solution of major problems. The divergence of opinions is even more marked in regard to the area of humoral sleep factors. While the emergence of new sleep factor candidates is considered by some as a breakthrough that will soon provide the ultimate remedy for insomnia, others deny *a priori* the possibility that a specific endogenous sleep-promoting substance may exist. To promote a positive, yet critical attitude towards research, it is of foremost importance that sleep researchers themselves define their terms and clarify their concepts. Recently, criteria have been proposed which a "sleep substance" should fulfill (*6, 16*). The main objective of this paper is the discussion of basic principles of sleep regulation which are relevant also for the study of humoral sleep-promoting compounds. Rat data will serve to illustrate various aspects of sleep regulation in mammals.

I. HOMEOSTATIC AND CIRCADIAN MECHANISMS OF SLEEP REGULATION

The 24-hr sleep/wake cycle of the rat shows two clearly delimited periods: in the 12-hr light phase the rat spends close to 80% of the time in sleep, whereas in the 12-hr dark phase it is close to 80% awake (Fig. 1, left top panel). Note in Fig. 1 the rapid transitions between the high and low sleep periods. The sleep/wake cycle persists with a close to 24-hr periodicity even in the absence of a light-dark cycle or other environmental synchronizers, and constitutes therefore a genuine circadian rhythm (4).

Traditionally, non-REM sleep (NREMS) and REM sleep (REMS) are discriminated as the two major substates of sleep. In the rat, REMS represents 15–20% of recording time in the light phase and declines to very low levels in the dark phase (Fig. 1). This sleep state is character-

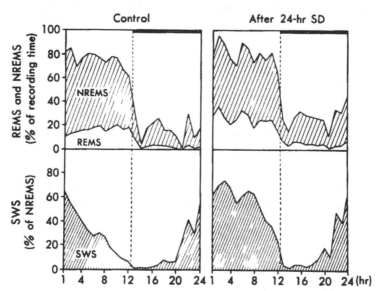

Fig. 1. Effect of 24-hr sleep deprivation (SD) on the sleep state distribution in the rat. The curves represent mean hourly values (n=6) expressed as percent of recording time (for NREMS and REMS) or as percent of NREMS (for slow wave sleep (SWS)). SWS was defined on the basis of the zero-crossing analysis as the NREMS fraction with the lowest 30 % of zero-crossing values on baseline days. The dark phase is indicated by a bar below the top abscissa and an interrupted vertical line (modified from ref. 5).

ized by a low-amplitude, high-frequency electroencephalogram (EEG) and a prominent theta-rhythm (6–9 Hz) of hippocampal origin which can be easily recorded by cortical surface electrodes. In man, substates of NREMS are usually discriminated on the basis of EEG amplitude and frequency criteria. They show a typical distribution across the night with slow wave sleep (SWS; stages 3+4) dominating in the first part, and "spindle sleep" (stage 2) in the second part. EEG frequency analysis revealed that also in the rat NREMS is an inhomogenous state, and that the temporal distribution of its substates resembles human sleep. Thus SWS as defined by EEG zero-crossing criteria (5, 7) represents more than 60% of NREMS in the first light-hour, and then progressively declines (Fig. 1, left bottom panel). In the dark phase SWS shows a gradual increase. However, due to the low level of sleep in this phase, the absolute amount of dark-time SWS remains small.

Sleep-deprivation (SD) is a potent tool for studying the regulatory aspects of sleep. In the rat forced locomotion has been used to enforce prolonged waking. The animal is kept in a slowly rotating cylinder where he has access to food and water (5). The procedure involves only a minimal amount of stress as indicated by the non-significant increase in plasma corticosterone (23). Moreover, experiments involving voluntary locomotion demonstrated that the effects of SD on sleep are not due to increased locomotion (13). Apart from brief episodes of EEG synchronization during motionless periods, no sleep signs were observed during SD (5).

A 24-hr SD period terminating at light onset has marked effects on the consecutive sleep period (Fig. 1, right panels): the amount of REMS is massively increased, while the amount of NREMS is reduced. Moreover, as shown in the lower right panel of Fig. 1, the substates of NREMS show an altered distribution. Thus a high level of SWS persists for several hours and declines only in the last third of the light phase. Also delayed effects of SD were observed. They include an increase of REMS (significant) and NREMS (non-significant) in the dark phase as well as a significant reduction of the EEG amplitude throughout the second post-SD day (5, 12, Trachsel et al. in preparation).

In the experiment illustrated in Fig. 1, recovery sleep occurred during the light phase, the rat's circadian sleep period. To study the interaction between the enhanced "sleep pressure" due to extended waking, and the circadian sleep/wake rhythm, the 24-hr SD period was made to terminate at dark onset (5). In this schedule a conflict arose between the high sleep propensity due to prior waking and the circadian tendency for waking and activity. Although a sleep rebound was seen during the first dark-hours of the recovery period, sleep remained below 50% of recording time and was followed by prolonged wake episodes in the second part of the night. SWS was enhanced in the first recovery hours and exhibited a delayed, second peak after light onset. These results demonstrate the potent influence of the circadian oscillator on sleep and waking. They also indicate that two separate processes determine sleep propensity: a homeostatic, sleep/wake-dependent process, and a circadian, sleep/wake-independent process. This view is supported by recent experiments in which the circadian rest-activity rhythm of rats was extensively disrupted or completely abolished by lesions of the suprachiasmatic nuclei (24). In these animals, a 24-hr SD period still induced a significant increase in REMS and SWS. The results demonstrate that the homeostatic component of sleep regulation is morphologically and functionally distinct from the circadian component. This conclusion is further supported by experiments in which rats maintained under continuous darkness were subjected to 24-hr SD (ref. 2: Fig. 3, and Borbély and Tobler, unpublished results). Neither the period nor the phase-position of the free-running circadian rest-activity rhythm was affected by the deprivation procedure.

Based on the evidence for two separate processes determining sleep propensity, a formal model of sleep regulation was established (2, 3, 10). The time-course of the sleep/wake-dependent process (Process S)

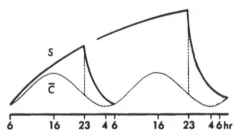

Fig. 2. The two-process model of sleep regulation. Time-course of sleep processes after regular (left) and extended wake period (right). Process S and the negative function of Process C (curve $\bar{C}$) are indicated. Ordinate: level of S in arbitrary units; abscissa: time of day (modified from ref. 3).

was estimated from the EEG slow wave activity values computed for successive NREMS/REMS cycles of human sleep (*8*). The decline under baseline conditions (Fig. 2, left part) could be approximated by an exponential function. After a continuous waking period of 40.5 hr, the initial slow wave activity values were increased by approximately 40%, but still showed an exponential decline. By interpolation of the initial and terminal sleep values, the rising portion of Process S was defined as another exponential function. Process C represents the circadian component of sleep propensity which is unaffected by sleep or waking. The phase-position of C was derived from the circadian rhythm of vigilance during prolonged waking (*3*). To facilitate the vizualization of the combined effect of the two processes, the negative function (mirror image) of Process C (designated by $\bar{C}$) has been plotted in Fig. 2. Curve $\bar{C}$ may be regarded as reflecting the "wake-up propensity" which is lowest when circadian sleep propensity is highest (*i.e.*, in the early morning hours). It should be also noted that curve $\bar{C}$ corresponds closely to the circadian rhythm of body temperature (see Ref. *3*). Total sleep propensity is represented by the difference between S and $\bar{C}$. In accordance with experimental data, sleep propensity is a nonmonotonic function of elapsed wake time. The model has been expanded to account also for the NREMS/REMS cycles. On the one hand, REMS propensity corresponds largely to the circadian component of sleep propensity. On the other hand, a reciprocal inhibitory interaction is assumed to occur between the NREMS and REMS controlling processes. In the model, these two factors are responsible for the cyclic alternation of the two substates of sleep as well as for their opposite trends across the night (*3*). These considerations may be applied also to animal sleep to account for the opposite trends of REMS and SWS (Fig. 1, left). Moreover, an extended version of the model can simulate the polycyclic pattern of rodent sleep (*10*).

II. REGULATION OF NONREM SLEEP AND REM SLEEP

In the studies mentioned, SWS was assessed on the basis of the EEG zero-crossing analysis (*5*), a technique that lends itself readily to the quantification of large amounts of data, but does not allow a precise measurement of the EEG frequency distribution. To overcome this limitation, we have applied in recent experiments the method of spectral analysis to study sleep regulation in the rat (*9*). Under baseline conditions, EEG slow wave activity in NREMS (*i.e.*, the spectral power density in the low frequency bands) exhibited a progressive decline throughout the first 8 hr of the light-phase. While the largest reduction in power density occurred in the 1.75–2.0 Hz band, the decrement was not limited to the delta range, but extended up to the frequencies in the theta range. Twenty-four hr SD caused a prominent enhancement of slow wave activity. While the progressive decline was still present after SD, the initial slow wave activity values were much higher than under baseline conditions.

Spectral analysis of the REMS EEG revealed unexpectedly a massive increase of power density in the theta band after SD. The maximum increase was seen in the first 2-hr period of recovery sleep, yet the values remained significantly elevated throughout the entire 8-hr recording period.

Are separate processes involved in the regulation of NREMS and REMS? We have previously argued that SWS could be regarded as an intensity dimension of NREMS which allows compensation for a sleep deficit without prolongation of sleep time (*2, 5*). In contrast to NREMS, a deficit in REMS would have to be compensated by a prolongation of REMS time, since no electrographic intensity indicator has been described. In the light of our recent findings, however, these arguments need to be modified, since theta activity may be considered as a REMS intensity indicator. This implies that a REM sleep deficit could lead to a compensatory increase in both time and intensity of this sleep state. However, there is evidence for more basic differences of the regulatory characteristics of REMS and NREMS: a moderate extension of wake time (12 hr in rat (*5*); 40 hr in man (*8*)) enhanced slow wave activity during recovery, but did not significantly affect the amount of REMS. Therefore, in contrast to NREMS, a compensatory response of REMS is elicited only by a severe deficit. Recent experiments in the rat support this interpretation (Tobler and Borbély, in preparation). Thus when the animals were sleep deprived for 3, 6, 12 or 24 hr and subsequently recorded during the first 8 hr of the light phase, slow wave activity in NREMS was enhanced as a function of prior wake time

(Fig. 3A). A significant increase occurred when the SD period exceeded 3 hr. In contrast, REMS time was not significantly affected by SD periods ranging from 3 hr to 12 hr, but was dramatically increased after 24-hr SD (Fig. 3B). Similarly, the theta activity in REMS was significantly enhanced after 24-hr SD, but not by the shorter SD-schedules.

When analyzing the regulatory principles underlying the substates of sleep, the question of how to define a "state" is of crucial importance. Results from our recent study (9) indicate that this is not a trivial problem. Thus, during the initial recovery period from 24-hr SD EEG slow wave activity was consistently observed in the awake and behaving animal (see ref. 9 for references of similar observations). Furthermore, the enhancement of theta activity by SD was not only present in REMS, but also during waking. This result supports the notion that active waking and "active sleep" (i.e., REMS) may be closely related in terms of the state controlling mechanisms. High "theta propensity" may manifest itself as active waking when the vigilance level is high, or as REMS when the vigilance level is low (see ref. 9). If theta activity is regarded as an indicator of overt or covert arousal, its enhancement by SD would reflect an activating effect of prolonged waking.

III. IMPLICATIONS FOR HUMORAL SLEEP FACTOR RESEARCH

If a specific endogenous substance is involved in sleep regulation, its administration should have effects that are comparable to those induced by physiological manipulations. As has been pointed out, the effect of SD is dependent on the circadian period in which recovery sleep is allowed to occur. The amount of total sleep is little affected, if the recovery from SD coincides with the circadian sleep period (5, 9). Inoué and coworkers (14) have recently reported that in contrast to the sleep-promoting action of nocturnal administration, diurnally administered sleep substance candidates (i.e., delta-sleep-inducing peptide (DSIP), prostaglandin D_2 (PGD_2), and uridine) did not significantly affect sleep time. In this respect, increasing the "sleep pressure" by humoral or physiological manipulations shows similar effects. Nevertheless, these observations alone constitute insufficient evidence for an involvement of the tested compounds in physiological sleep regulation, since also the susceptibility to pharmacological agents exhibits a circadian rhythmicity (*19*).

The enhancement of "sleep pressure" by SD in donor animals has frequently been used to increase the level of a putative sleep factor. As has been pointed out, the increase in EEG slow wave activity represents one of the most sensitive responses to sleep loss. Various sleep substance candidates and their derivatives such as factor S (*20*), muramyl peptides (*17*), and interleukin-1 (*18, 25*) were reported to enhance SWS. However, in most of these studies, the spectral frequency distribution and its time-course have not been analyzed and compared to the physiological changes after SD.

The problem of whether the substates of sleep are regulated by separate mechanisms is most relevant for the study of humoral sleep factors. We have seen that a moderate SD period leads to an increase in SWS alone, whereas a more severe sleep deficit results in a rebound of both SWS and REMS (**Fig. 3**). Various procedures have been applied to achieve a selective REMS deprivation which was followed by

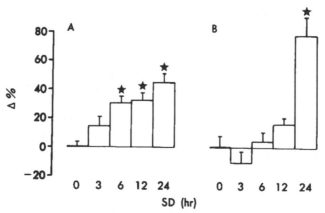

Fig. 3. Effect of sleep deprivation on slow wave activity in NREMS (A) and on REMS time (B). Mean values (*n*=6) with S.E.M. are shown for the first 8 hr of the light phase. The bars indicate deviations from baseline (=0 hr deprivation) plotted as a percentage of baseline (=100 %) for various durations of sleep deprivation. Slow wave activity was obtained by integrating the standardized NREMS spectra over the entire frequency range (0.75–25.0 Hz). The baseline REMS value corresponds to 13.2 % of recording time. Asterisks indicate significant difference from baseline ($p < 0.05$; two-sided Wilcoxon matched pairs signed ranks test).

a REMS rebound. However, since the effect on EEG slow wave activity has usually not been analyzed in detail, the degree of selectivity is still an unresolved problem. In a recent study, sleep substance candidates and their derivatives were reported to significantly enhance either NREMS alone (*e.g.*, DSIP, muramyl dipeptide (MDP), and PGD_2) or both NREMS and REMS (SPS-B and uridine) (*15*, see Chapter 1). It is still unclear whether this selectivity reflects basic differences in the mechanisms of action, or if it is due to less specific factors such as dose or route of administration.

Two groups have used selective REMS deprivation to see whether a REMS-controlling factor accumulates in the cerebrospinal fluid (CSF). Sallanon and coworkers (*22*) used donor cats that had been REMS-deprived for 17 hr. The recipient cats were pretreated with *p*-chlorophenylalanine (PCPA) to induce insomnia. Intracerebroventricular (i.c.v.) injection of CSF from donors enhanced both REMS and NREMS. Adrien and Dugovic (*1*) used REMS-deprived rats as donors, and rats pretreated with α-methyldopa or propranolol as recipients. I.c.v. infusion of CSF from donors restored REMS that had been suppressed by the pretreatment. When assessing these interesting results, it is important to keep in mind that REMS enhancement was not reported for untreated animals.

It is one of the basic assumptions of the SD approach that putative sleep substances accumulate during waking. Therefore the level of the substance in the donor animal should be a function of the duration of prior waking. The results of two studies confirm this prediction: the suppression of motor activity of rats receiving goat CSF depended upon the duration of SD in the donor animal (*11*). Thus activity was increasingly reduced as donor CSF obtained after 0-hr, 12-hr, and 24-hr SD was administered. CSF after 72-hr SD caused no further suppression of activity. Adrien and Dugovic (*1*) deprived their donor rats of REMS for 1, 2, 3 or 4 days. The longer the deprivation lasted in the donor, the more REMS was restored in the pretreated recipient rats (see Chapter 19). Such a time-dependence of the endogenous level of a putative sleep factor is one of the most compelling arguments for its involvement in the homeostatic mechanism of sleep regulation.

When investigating the presence of endogenous sleep factors, one should keep in mind that endogenous wake-promoting compounds could equally exist. Thus, after administration of a protein-free filtrate of goat CSF, Fencl and coworkers (*11*) observed a dose-dependent, long-lasting hyperactivity in the recipient rats. Further indications of an activity-promoting CSF factor were obtained by Sachs and coworkers (*21*) who observed an enhancement of light-time motor activity in recipient rats after i.c.v. infusion of CSF obtained from donor rats during their circadian wake period.

In conclusion, it is essential to evaluate effects of sleep substance candidates in the context of physiological sleep regulation. Careful behavioral observations; long-term recordings encompassing several 24-hr cycles; a detailed EEG analysis in regard to frequency and time-course; and the measurement of relevant correlates of vigilance states and circadian rhythms (*e.g.*, body temperature) are prerequisites for a meaningful interpretation of the effects of humoral factors. A major limitation in testing putative sleep substances derives from the difficulty of administering a compound in physiological concentration to the cerebral target area without inducing non-specific effects due to stress or the infusion procedure *per se*. The technique of long-term infusion with concomitant sleep recording developed by Inoué and coworkers (*15*) constitutes an important step towards a physiological approach of humoral sleep regulation.

SUMMARY

A homeostatic and a circadian aspect of sleep regulation can be discriminated. The former is reflected by the increase in SWS as a function of prior waking, whereas the latter is evident from the free-running circadian sleep/wake rhythm in the absence of time cues. Abolition of the circadian rhythm does not disrupt sleep homeostasis. The two-process model of sleep regulation is based on the interaction of a homeostatic and a circadian process. The two substates of sleep show different regulatory characteristics. While even a moderate sleep deficit gives rise to an increase in SWS, a more severe deficit is required for a REMS rebound. There are indications that REMS may be related to active waking. Putative humoral sleep factors should be evaluated in the context of physiological sleep regulation.

Acknowledgment

The studies presented in this paper were supported by the Swiss National Science Foundation, grants 3.171-0.81 and 3.518-0.83.

REFERENCES

1 Adrien, J. and Dugovic, C. (1985). This volume, pp. 227–236.
2 Borbély, A.A. (1982). In *Current Topics in Neuroendocrinology*, ed. Ganten, D. and Pfaff, D., vol. 1, pp. 83–103. Berlin: Springer-Verlag.
3 Borbély, A.A. (1982). *Hum. Neurobiol.* **1**, 195–204.
4 Borbély, A.A. and Neuhaus, H.U. (1978). *J. Comp. Physiol.* **128**, 37–46.
5 Borbély, A.A. and Neuhaus, H.U. (1979). *J. Comp. Physiol.* **133**, 71–87.
6 Borbély, A.A. and Tobler, I. (1980). *Trends Pharmacol. Sci.* **1**, 356–358.
7 Borbély, A.A., Neuhaus, H.U., and Tobler, I. (1981). *Behav. Brain Res.* **2**, 1–22.
8 Borbély, A.A., Baumann, F., Brandeis, D., Strauch, I., and Lehmann, D. (1981). *Electroenceph. Clin. Neurophysiol.* **51**, 483–493.
9 Borbély, A.A., Tobler, I., and Hanagasioglu, M. (1984). *Behav. Brain Res.* **14**, 171–182.
10 Daan, S., Beersma, D.G.M., and Borbély, A.A. (1984). *Am. J. Physiol.* **246**, R161–R178.
11 Fencl, V., Koski, G., and Pappenheimer, J.R. (1971). *J. Physiol.* **216**, 565–589.
12 Friedman, L., Bergmann, B.M., and Rechtschaffen, A. (1979). *Sleep* **1**, 369–391.
13 Hanagasioglu, M. and Borbély, A.A. (1982). *Behav. Brain Res.* **4**, 359–368.
14 Inoué, S., Honda, K., Komoda, Y., Uchizono, K., Ueno, R., and Hayaishi, O. (1984). *Neurosci. Lett.* **49**, 207–211.
15 Inoué, S., Honda, K., Komoda, Y., Uchizono, K., Ueno, R., and Hayaishi, O. (1984). *Proc. Natl. Acad. Sci. U.S.* **81**, 6240–6244.
16 Jouvet, M. (1984). *Exp. Brain Res.* (Suppl. 8), 81–94.
17 Krueger, J.M., Walter, J., Karnovsky, M.L., Chedid, L., Choay, J.P., Lefrancher, P., and Lederer, E. (1984). *J. Exp. Med.* **159**, 68–76.
18 Krueger, J.M., Walter, J., Dinarello, C.A., Wolff, S.M., and Chedid, L. (1984). *Am. J. Physiol.* **246**, R994–R999.
19 Lemmer, B. (1984). *Chronopharmakologie. Tagesrhythmen und Arzneimittelwirkung.* Stuttgart: Wissenschaftliche Verlagsgesellschaft.
20 Pappenheimer, J.R., Koski, G., Fencl, V., Karnovsky, M.L., and Krueger, J.M. (1975). *J. Neurophysiol.* **38**, 1299–1311.
21 Sachs, J., Ungar, J., Waser, P.G., and Borbély, A.A. (1976). *Neurosci. Lett.* **2**, 83–86.
22 Sallanon, M., Buda, C., Janin, M., and Jouvet, M. (1981). *C. R. Acad. Sci. Paris* **292**, 113–117.
23 Tobler, I., Murison, R., Ursin, R., Ursin, H., and Borbély, A.A. (1983). *Neurosci. Lett.* **35**, 297–300.
24 Tobler, I., Borbély, A.A., and Groos, G. (1984). *Neurosci. Lett.* **42**, 49–54.
25 Tobler, I., Borbély, A.A., Schwyzer, M., and Fontana, A. (1984). *Eur. J. Pharmacol.* **104**, 191–192.

3

TISSUE RESTITUTION AND SLEEP, WITH PARTICULAR REFERENCE TO HUMAN SLOW WAVE SLEEP

JAMES A. HORNE

Department of Human Sciences, Loughborough University, Loughborough, Leicestershire LE11 3TU, England

It is often assumed that sleep is a restorative state (*i.e.*, enhanced anabolism and tissue repair following the "wear and tear" of wakefulness). The theory has come first, and then evidence favouring this viewpoint (*e.g.*, the sleep-related human growth hormone (hGH) output, and increases in mitosis during sleep) is interpreted accordingly, from an a posteriori basis. Such an approach is contrary to the usual scientific method where the null hypothesis should hold (*i.e.*, sleep is not restitutive) until otherwise proven. The argument will be made here that the evidence so far does not lead us to reject the null hypothesis, at least for the majority of organs. However, the forebrain may be the exception, and for this reason the ensuing discussion will separate body from brain, so that the term "body" specifically excludes the brain.

For the body, the primary stimulus for restitution is food (amino acid) absorption from the gut, coupled with physical inactivity (*e.g.*, sitting). In those mammals, such as rodents, which are unable to display any prolonged periods of relaxed wakefulness, sleep is the immobiliser, and consequently, heightened restitution may be associated with sleep because of the inactivity rather than because of sleep itself (24–26). But for humans and other mammals with more developed forebrains, where there is the behavioural repertoire to display relaxed wakefulness (an advanced form of behaviour, 9), body restitution is quite able to occur during wakefulness. In fact, because we humans usually cease eating by about 20:00 each day, and do not re-feed until about 12 hr later, sleep is a fasting state, where tissue degradation occurs (8, 58), and it is more likely that the sleep-hGH release may be associated with the slowing up of protein breakdown and oxidation (and the mobilisation of fats as an alternative energy substrate), not with increased anabolism. This topic will be returned to in Section I-2.

Apart from the brain, and as far as it is known, no body organ enters any unique and essential physiological state during sleep. A human subject lying relaxed but awake in a darkened and sound dampened room can attain levels of physical relaxation and metabolic rate similar to those of sleep (54). However, the cerebrum remains in a condition of "quiet readiness," prepared to respond to any stimulus. Only during sleep (particularly during that form of sleep exhibiting delta electroencephalographic (EEG) activity, i.e., stage 4 sleep, and to a lesser extent during stage 3 sleep—collectively called human slow wave sleep: hSWS) is there some release for the cerebrum from this high state of vigilance. But for most mid and hindbrain structures sleep is not a condition of rest, as, for example, all vital functions still have to be maintained. There are signs that the cerebrum is in a unique state during hSWS, which could be indicative of some form of off-line recovery, but because of technological and obvious ethical limitations there is as yet no direct evidence of any restitutive event (e.g., dendritic growth, consolidation of memory processes, etc.). If the cerebrum does need off-line recovery for whatever purposes, together with reduced neuronal firing rates and isolation from sensory input, then of all the sleep stages, hSWS is the best fit for these criteria. It should be noted that of all the sleep stages, hSWS is the most positively correlated with the duration of prior wakefulness (60), and in this respect would seem to be the most appropriate for any brain restitutive role.

Indications that the human cerebrum enters a peculiar state of isolation during hSWS come from a variety of studies (5, 51, 57). A report on the evocation of epilepsy during sleep (*51*) concluded that during hSWS there was a "functional disconnection of the cortex," not found in other sleep states. Another study (*57*), on evoked potentials during sleep, found that the blocking effect of hSWS was unique, and recently (*5*), it has been shown that hSWS reflects a partial disconnection between the hemispheres.

Human SWS, or at least the first 2–3 sleep cycles containing hSWS, seems to be an obligatory form of sleep (*17, 21, 24*). Sleep reduction studies (see ref. *38* for review) all show that the reductions are at the expense of all stages except hSWS, for sleep lengths down to 5.5 hr-sleep per night (from 7.5 hr). Age-matched natural short (less than 6.5 hr) and long (over 8.5 hr) sleepers have similar absolute levels of hSWS (*59*), and so it seems do naturally occurring very short sleepers (less than 3 hr per day) when their sleep is compared with age-matched controls (*24*). It seems that these very short sleepers (*e.g.*, *39*) have the normal first two sleep cycles and then dispense with the rest of sleep. Only about 30% of the total amount of sleep lost during sleep deprivation seems to be made up, and it appears to be the quality of this sleep which counts. Calculations (*24*) from data of 2–11 days sleep deprivation show that about 80% of the lost stage 4 sleep is reclaimed, with about 50% of the lost REM sleep. None of the lost stages 1 and 2 sleep are reclaimed.

These findings concerning the apparent need for hSWS have led me (*17, 21, 22, 24*) to conclude that the first three sleep cycles containing hSWS and the associated REM sleep represent an obligatory form of sleep which gives way to a more facultative sleep as the night progresses. This latter form of sleep can be reduced or extended according to the needs of the organism, particularly, safety, boredom, energy conservation, and circadian factors. This concept of obligatory and facultative sleep, which is not new (*50*), applies to other mammals, as they also seem to have analogues of hSWS during the first part of sleep. Humans seem to have the capacity not only to reduce daily sleep by up to 1.5–2.0 hr without adverse effects such as increased daytime sleepiness (*38*), but can extend sleep by a similar amount (*e.g.*, *11*). Similarities can be drawn here with food intake. Most people are able to reduce or increase this by about 20% for long periods, respectively by making more efficient use of foodstuffs or by dumping heat energy and nitrogen, without adverse effects on health, but simply with a resetting of the bodyweight "setpoint."

I. BODY RESTITUTION AND SLEEP

1. Sleep Deprivation

For many proponents of a body restitutive role for sleep, the type of sleep which is considered to be specifically oriented towards this restitution is hSWS, and REM sleep is often seen to be for brain restitution. I have argued against both these viewpoints (17, 21–26, 28). However, this discussion will not concern itself with REM sleep, (see ref. 28 for a recent review on REM sleep). The sleep-related hGH output is particularly associated with hSWS, and the peaks of mitosis are concomitant with hSWS. These are two of the main reasons for the postulate that it is hSWS that is related to body restitution. Also, there are the studies on the effects of exercise on subsequent sleep, which find that of all the sleep characteristics that could alter following exercise (presumably by increasing "wear and tear"), it is hSWS which changes, by increasing (see ref. 23 for review). This outcome, which is misleading, adds to the "evidence" supporting a body restitutive role for hSWS. These issues will be debated further in section I-3.

The obvious method for assessing a body restitutive role for sleep is to see what happens during sleep deprivation. Whilst there have been few studies of hSWS deprivation per se, there have been many of total sleep deprivation (TSD), of up to 11 days without sleep. Reviews of the literature (3, 19, 43, 61) all report how uneventful TSD seems to be for body functioning, although there are obvious behavioural changes pointing to central nervous system (CNS) impairment (and perhaps some essential need for the brain to sleep). Hormonal profiles and assessments of other urinary and blood constituents of TSD subjects strongly indicate a general deactivation rather than any alarm response (3), unless subjects are additionally stressed, for example, through apprehension (19, 26). As far as measures have permitted, there seems to be no significant change in protein metabolism during TSD (19, 26), which would be expected if sleep were the primary condition for anabolism (see below). If the voluntary muscles of TSD subjects were degenerating in any way through "lack of anabolism," or the energy transfer processes were becoming impaired, then the physiological efficiency to do physical work would be reduced, and for example, there would be adverse changes to the rates of oxygen uptake and carbon dioxide production at fixed workloads. But this is not the case (19, 32), at least up to 3–5 days TSD (the longest of this type). Exercise endurance may decline, but this seems to be through psychological rather than physiological mechanisms (see ref. 32 for review).

The response of certain body organs or systems to TSD still remains relatively unexplored. There are indications that the immune system changes its characteristics in ways that do not seem to be of much clinical significance (44). Recent findings with respiratory chemosensitivity indicate that one night of TSD diminishes this response with some form of impairment to the CNS control mechanisms (26). Also, respiration and heart rates become more irregular (20), but for reasons which are not clear. Another CNS regulating mechanism which seems to be affected by TSD is thermoregulation (19, 26), but not to any serious extent with humans. Subjects feel the cold more and there is usually a small drop in body temperature of about 0.5°C (19, 26, 33), although the circadian rhythm of body temperature is maintained (3). Subjects compensate by putting on additional clothing, and for this reason the extent of the thermoregulatory change has never been really established, except in a recent investigation (52) of thermoregulation during exercise following one night of TSD. Here, there was a significant fall in evaporative heat loss.

The recent Chicago Sleep Laboratory findings (e.g., 6, 49) with TSD in the rat have concluded that thermoregulation and energy balance are impaired. But in such a small mammal with a relatively high amount of surface area in proportion to body volume, the consequences are far more serious than for humans, particularly when the animal is prevented from invoking behavioural countermeasures (as was the case in the Chicago studies). In fact, the Chicago Laboratory indicates that impaired thermoregulation was the primary cause of death of their animals, despite a very high food intake. Careful histopathological investigations of the major organs by these researchers revealed nothing remarkable when the TSD animals were compared with a stress control group. Both groups seemed to display equal amounts of pathology, even for the CNS, where a shrinkage of neuronal perikarya and vacuolisation were reported. However, such CNS analyses are relatively crude, and give little indication of possible neurological disorder (neurological tests were not performed on the animals).

For human TSD subjects, standard neurological assessments have not revealed anything of great clinical significance apart from cerebral impairment. Although there are clear neurological changes such as ptosis and some hand tremor, these are only minor, even in the longest studies (see refs. *17* and *26* for reviews). Diplopia and problems with visual accommodation are common (*15*). The EEG changes are indicative of heightened sleepiness (increased theta activity and decreased alpha). But in some subjects, particularly those with some history of paroxysmal EEG activity, there is a significant risk of such activity occurring, even after only one night of TSD (*17, 26*). Apart from sleepiness, behavioural changes include irritability, suspiciousness and visual "illusions," but seldom true hallucinations (*17, 26, 46*). The manifestation of these phenomena does depend considerably on the personality to start with, and on the rapport with the experimenter and other subjects. The major effect of sleepiness is on motivation (*27, 33, 43, 61*), and in order to keep subjects relatively alert, constant stimulation is required. There is a clear circadian rhythm of sleepiness during TSD, with the effects being worst in the early morning and after lunch, and least during the early to late evening period. Of particular interest is that these behavioural and psychological effects of TSD seem to be largely reversed after only one recovery night of sleep, even for the longer studies (*46*). Such sleep contains relatively large amounts of stage 4 sleep (*16, 40*).

2. *Protein Synthesis, Mitosis, and hGH Output in Sleep*

Despite the unremarkable outcome from human TSD studies with regard to body restitution, there are still findings during sleep *per se* suggesting that sleep stimulates increases in protein synthesis and cell division (*i.e.*, it is restorative). These arguments have been compiled by Adam and Oswald (*1, 2*). Most of their evidence comes from investigations with rodents, and there is probably little doubt from the variety of literature they cite that sleep in these animals is associated with these body restitutive events. However, I have argued (*19, 22,*

24–26) that the key to these increases in protein synthesis and mitosis in sleep is increased food absorption, physical inactivity, and to some extent, lowered corticosteroid output, all of which are concomitant with sleep, not caused by sleep. In rodents, food absorption peaks are delayed to the end of the wakefulness period, mostly because of slowing down of absorption through physical activity. As noted in the Introduction, for these animals sleep is the immobiliser. Also, it should be remembered that in rodents the main influence on 24-hr corticosteroid rhythms is not sleep, but feeding (*4*). So, for rodents, sleep is only a "vehicle" for increased body restitution. I have elaborated elsewhere (*24, 26*) the crucial role physical rest has in facilitating increases in mitosis in rodents.

Nearly all of the studies of protein synthesis changes over 24 hr in rodents, cited by Adam and Oswald (*1, 2*), only measure protein synthesis and not other forms of protein metabolism. Protein is unstable, and in all cells is in a state of flux, being not only synthesised but also broken down into the constituent amino acids. So simply measuring synthesis will not determine whether cell restitution and growth is actually occurring (although it so happens that at sleep onset in rodents, this is the case). For such a conclusion, breakdown rates must be measured to see whether synthesis is exceeding breakdown. It is quite possible for synthesis to increase and for breakdown to occur at a higher rate, leading to a net decrease in cell protein content. Measures of protein synthesis and breakdown have been made over 24 hr in humans (*8*), and it is clear that food absorption is the main stimulus (*58*) to increased protein synthesis and to a decrease in protein breakdown, *i.e.*, causing an increase in cell protein content. Physical rest (*e.g.*, sitting) facilitates this. Both food absorption and physical rest usually occur for a few hours after the main meals of the day (particularly for the evening meal) in humans, and for most of us the rise in protein content of tissue is completed before the sleep period. Consequently, sleep becomes a fasting state (*8, 58*) for most people, and a condition of net protein loss from cells because protein breakdown exceeds synthesis. One of the many functions of hGH (*22, 45*) is to spare protein against increased rates of breakdown which can occur during fasting. Otherwise, this protein could be used as an energy substrate. Instead, hGH mobilises fats for this purpose. Nighttime sleep in humans is relatively long as mammalian sleep goes, and this, coupled with the fasting state, has led me to suggest (22, 24–26) that the sleep-hGH release may be an anticipatory response to a further and relatively long period (i.e., 8 hr) without food. It is interesting to note that one of the most potent stimuli to suppress the sleep-hGH release is an elevation in blood lipids (see ref. 45 for review).

The circadian peaks in mitosis which occur in the sleep period in humans are simply a time of day phenomenon, as they are apparent at this time in the absence of sleep (53). Clearly, they have nothing to do with the sleep-related hGH output which is usually concomitant with these peaks, as of course, this output is absent if the subject remains awake. Exercise impairs mitosis (12), but light activity, including walking seems to create no impediment in this respect (12, 53). In those cells that can divide, mitosis usually follows an increase in cell size (G2 phase), which is in turn affected by the protein content of cells. Hence, food absorption modulates mitosis indirectly. The final factor which affects mitosis in humans, corticosteroid output, has a circadian rhythm independent of sleep and wakefulness (4, 45) and, for example, continues unabated during sleep deprivation (3, 4, 19, 26), further evidence demonstrating the independence of mitosis peaks from sleep.

Another factor which must be appreciated when considering the functional significance of the sleep-related hGH output, is that it is not a common event amongst mammals. It is not apparent in the cat and dog, unless there is prior sleep deprivation (56). It is absent in the rhesus monkey, and for the baboon there are conflicting reports (see ref. 48 for review). This variation amongst mammals does raise doubts about any restitutive role this hormone might have for sleep, and is a matter not often addressed by supporters of this role. With regard to the suggested possible protein sparing role of hGH during sleep, and its absence in the sleep of the cat and dog, one must remember that for "gorging" carnivores which depend on protein as the primary energy substrate, such a function for hGH would be less appropriate than is the case for the omnivorous human.

A final comment should be made about the hypothesis of Adam and Oswald (1, 2) concerning putative increases in cellular energy charge—EC (i.e., adenylate ratios) during sleep, and how such increases are supposed to promote anabolism. This is a very misleading issue, which I have covered in detail elsewhere (24–26). In essence: i) there is no evidence that EC increases from waking to sleep in humans, and for rodents the findings are doubtful, ii) the evidence implying a positive correlation between EC and anabolism is taken from extreme and often life threatening examples (26) irrelevant to the natural states of sleep and wakefulness, iii) the half life of ATP is in seconds, and in an average day a human will turnover at least one body weight of ATP, which enables maximal levels to be maintained quite easily, unless anaerobic exercise is performed (13).

3. Exercise, Heating and Subsequent hSWS

The EEG is only a guide to cerebral activity, and is a poor indicator of body functioning. Despite this limitation of the EEG, many sleep researchers see the hSWS changes which can occur following daytime exercise in the light of body restitutive events, and not in terms of how the cerebrum itself could be affected. Also, of course, the grounds for assuming that human sleep is restitutive for the body are doubtful. No exercise and sleep study has made a direct assessment of any body restitutive event (23).

Findings of increased hSWS following exercise are limited to highly physically trained ("fit") subjects (see ref. 23 for review). The relatively large number of studies on untrained ("unfit") individuals point to no hSWS effect, or to any other significant change to sleep. Even careful analysis of delta activity in stage 2 sleep reveals no substantive change (34) here. These contrasting hSWS findings between fit and unfit subjects pose problems for body restitution proposals for hSWS, as it could be argued that unfit subjects would be experiencing as much, if not more, muscle "wear and tear" during exercise when compared with fit individuals. I believe (23) that a major factor underlying the difference in hSWS response between these two subject types relates to the increased exercise endurance of fit subjects, and consequently, the greater likelihood of a higher and more sustained thermal load on both body and brain (exercise results in a core temperature rise proportional to the exercise load). Subsequent support for this proposal came from a study of ours (36) on fit subjects, which passively increased core temperature (through partial immersion of the subject in warm water) by an amount similar to that of a high intensity exercise condition. Both conditions lasted for 90 min, in the afternoon. A third condition controlling for exercise load, at half the rate for twice the time, produced only a small rise in core temperature. The first two conditions resulted in similar hSWS increases, with no change to hSWS for the third condition. It was concluded that the body heating effect of exercise is critical to the hSWS rise. Exercise alone is not the key factor.

A second study (*30*) repeated the high intensity exercise condition, again using fit subjects, but in a different control condition there was forced cooling of subjects undergoing identical exercise. In particular, there was facial cooling (which selectively reduces brain temperature, *7*). The exercise-related core temperature rise was reduced by half, and tympanic temperature measurement (an index of brain temperature) indicated a further temperature decrease. Whereas hSWS rose in the former condition to a level similar to that of the high intensity exercise in the previous study (*36*), the cool exercise resulted in no hSWS increase, again pointing to a key role for daytime thermal load, with exercise only acting as a vehicle for this effect.

Although unfit subjects do not encounter high thermal loads during exercise, owing to their limited exercise capacity, they can endure this heating through the passive method. Another study in our laboratory (*35*) gave the same passive heating to unfit subjects as that given to the fit subjects in the earlier investigation (*36*). This also resulted in a similar hSWS rise. A control condition of sitting in tepid water, to produce no core temperature change, had no effect on sleep. Interestingly, the passive heating produced a marked rise in sleepiness during the evening. It should be noted that core temperature increases produced by the various experimental conditions (*30, 35, 36*) all fell to baseline levels within an hour of ending the condition. The reasons why raised daytime body and/or brain temperature should increase hSWS are not clear, but several mechanisms are possible. A productive avenue of enquiry relates to the brain temperature increase, which will heighten brain neurochemical and metabolic processes at the time. It may be through these latter mechanisms that an increase in subsequent hSWS develops.

For example, the build up of a sleep-promoting substance might be associated with waking brain metabolism; increasing this metabolism through heating may enhance the accumulation of the substance. Another possibility is a rise in brain prostaglandin levels which, in the rodent at least, can be increased through raised brain temperature. A positive relationship between brain prostaglandin D_2 levels and the enhancement of non-REM sleep in rodents is described by Ueno *et al.* (see Chapter 16). Of perhaps further interest in this respect is the potent suppression of prostaglandin synthesis by acetyl salicylic acid (aspirin) (although the extent of this effect in the human brain is unknown). We have found (31) that 1,800 mg aspirin per day significantly reduces hSWS, and disrupts the organisation of sleep for several days following cessation of drug administration. Although the prostaglandin route is a possibility, owing to the long action of aspirin on prostaglandin synthesis, there are other mechanisms by which aspirin could have this effect (31), for example, through interference with tryptophan uptake by the brain. However, interference with this pathway in humans may affect REM sleep and not hSWS (see ref. 47 for further information). I suspect though, that prostaglandins are only one of perhaps many moderators of hSWS activity, and that we should explore deeper underlying mechanisms. Further clues to these mechanisms are discussed in the next section.

II. WAKING AROUSAL LEVELS AND SUBSEQUENT hSWS

It must again be emphasised that the source of the delta EEG activity in hSWS is the cerebrum, with sub-cortical structures "permitting" this activity to occur. I have suggested here, and elsewhere (17, 21, 22, 24, 37), that hSWS is influenced by the accumulation during wakefulness of some requirement within the forebrain, which may be of a restitutive nature. This is not a new concept, and stems from Feinberg's (10) earlier proposals. Although Webb and Agnew (60) avoided such a level of speculation, they demonstrated a clear relationship between the length of wakefulness and the accumulation of stage 4 sleep. I do not know what this waking requirement is (could it be related to a need for dendritic or glial growth and repair?). Waking neuronal and glial activities are manifested in the cerebral metabolic rate (CMR), and I will use this term as the general index of waking cerebral activity. During wakefulness, CMR runs at near maximal levels (55), even during relaxed states, probably because of the need to maintain vigilance (and "quiet readiness"—see p. 26). Consequently, unless there are more unusual factors, such as increased brain temperature or sustained arousal (see below), the "accumulated" CMR of wakefulness will remain fairly constant from day to day, and so does nightly hSWS. Increased behavioural arousal and attention, as well as heightened emotion, only result in a small increase in CMR (55) (much less than that for a 1.5–2.0°C rise in brain temperature found, for example, in our exercise studies, 30, 35, 36). However, if such aroused states persist for most of wakefulness, then the accumulated effect will be more consequential for overall CMR. This form of explanation seems to be the most parsimonious for our findings (16, 29, 37) of hSWS increases following sustained daytime attention under behaviourally demanding conditions. Increased sensory stimulation, particularly visual stimulation of a varied and complex nature, seems very effective in producing a hSWS increase. Humans are visually dominated mammals, with about one third of all sensory input coming from the eyes. Hence vision can occupy much of our attention.

There are many examples of positive correlations between waking CMR and subsequent hSWS. For example, in human ontogeny CMR falls substantially during the first few years of life, and so does hSWS. Both reach minimum levels in old age. But of course, this and other associations, such as increases in hSWS from hypothyroidism to euthyroidism, and then to hyperthyroidism (which result in successive increases of brain and body temperature and metabolism), do not point to causation.

From data on the rhesus monkey it seems that in non-REM sleep, CMR is about 20% lower than that of waking levels (42) (such studies have not been performed on humans). This could help explain why a hSWS need does not appear to build up in sleep, as it does in wakefulness (unless sleep is extended—see below). But CMR in REM sleep is at waking levels, or even slightly higher, and in this respect may contribute to a subsequent hSWS need. This may in part be responsible for a hSWS rebound at the end of very extended sleep (14). If the average young adult exhibits approximately 80 min of hSWS following about 16 hr of wakefulness (this proportion also applies to the extended sleepers, 14), then there is about 5.0 min hSWS per hour of wakefulness. This is a rough estimate, and does not take into account individual differences (particularly long and short sleepers), but is just to illustrate the next point. In a normal sleep length, the amount of REM sleep following the third non-REM cycle (usually the last to contain obvious amounts of hSWS) is about 70 min, which would, to follow this logic through, only require a little (6 min) subsequent hSWS, which can easily be overlooked. However, taking the findings from this recent 12–15 hr extended sleep study (*14*), a total of 203 min of wakefulness+ REM sleep followed after the first 6 hr of sleep, prior to the reappearance of hSWS. This would require about 18 min of hSWS on the 5.0 min hSWS per hour of wakefulness+REM sleep estimate, and compares favourably with the 21.5 min found.

My final point concerns the consequences of such extra sleep, as the amount of ensuing wakefulness up to the regular bedtime is obviously reduced. This seems to be the reason for the hSWS reduction in the subsequent sleep (*11*) (it is to be remembered that hSWS only manifests mild circadian influences, *60*). These findings, together with various nap studies (*e.g.*, *41*) which demonstrate that hSWS taken in a nap can be deducted from the subsequent night's hSWS quota with relatively little change to total sleep time, further indicate that hSWS can be manipulated relatively independently of total sleep length. This again points to two possible mechanisms controlling sleep, which I have described as obligatory sleep (mostly hSWS) and facultative sleep (total sleep time). In relation to the model proposed by Borbély (see this volume), I have, in effect, subdivided his "process *S*" into the two components, obligatory and facultative sleep.

SUMMARY

Stages 3+4 of human sleep (hSWS) seem to be an obligatory form of sleep. Although hSWS is claimed to be a condition of heightened tissue anabolism ("restitution"), it is argued here that, for the majority of tissues, the supportive evidence (*e.g.*, sleep-hGH output, mitotic increases, exercise effects on hSWS) is open to alternative, non-restitutive explanations which are more parsimonious. Rodent studies can be particularly misleading in this respect. The main stimulus to anabolism is food absorption, which is maximal in wakefulness. For most humans, sleep is a fasting condition of tissue degradation. Human and animal sleep deprivation findings do not support the restitutive hypothesis, except in the case of the brain, particularly the cerebrum, perhaps because this organ is unable to go "off line" during wakefulness. hSWS is a manifestation of cerebral functioning and is not an index of restitution for the rest of the body. hSWS reflects a unique form of cerebral shutdown not found in other sleep stages. This, and its high positive correlation with length of wakefulness, make hSWS the most likely candidate for "off line" cerebral restitution, if such a requirement is necessary. Waking cerebral metabolic rates may be a key to subsequent hSWS levels, and this may be a mechanism behind hSWS increases following daytime brain temperature rises (*e.g.*, *via* exercise), or sustained behavioural arousal. REM sleep may have similar effects on hSWS.

REFERENCES

1 Adam, K. and Oswald, I. (1977). *J. Royal Coll. Phys.* 11, 376–378.
2 Adam, K. and Oswald, I. (1983). *Clin. Sci.* 65, 561–567.
3 Åkerstedt, T. (1979). *Acta Physiol. Scand.* (Suppl.), 469.
4 Aschoff, J. (1979). In *Endocrine Rhythms*, ed. Krieger, D.T., pp. 1–61. New York: Raven Press.
5 Banquet, J.P. (1983). *Electroenceph. Clin. Neurophysiol.* 55, 51–59.
6 Bergmann, B., Kushida, C., Hennessy, C., Winterer, J., and Rechtschaffen, A. (1984). *Sleep Res.* 13, 185.
7 Cabanac, M. and Caputa, M. (1979). *J. Physiol.* 286, 255–264.
8 Clugston, G.A. and Garlick, P.J. (1982). *Hum. Nutr. Clin. Nutr.* 36C, 57–70.
9 Eisenberg, J.F. (1981). *The Mammalian Radiations*. London: Athlone Press.
10 Feinberg, I. (1974). *J. Psychiat. Res.* 10, 283–306.
11 Feinberg, I., Fein, G., and Floyd, T.C. (1980). *Electroenceph. Clin. Neurophysiol.* 50, 467–476.
12 Fisher, L.B. (1968). *Br. J. Dermatol.* 80, 75–80.
13 Flatt, J.P. (1978). In *Recent Advances in Obesity Research*, ed. Bray, G.A., pp. 211–228. New York: Newman Publishing.
14 Gagnon, P. and De Koninck, J. (1984). *Electroenceph. Clin. Neurophysiol.* 58, 155–157.
15 Horne, J.A. (1975). *Biol. Psychol.* 3, 309–319.
16 Horne, J.A. (1976). *Biol. Psychol.* 4, 107–118.
17 Horne, J.A. (1976). *Bull. Br. Psychol. Soc.* 29, 74–79.
18 Horne, J.A. (1977). *Physiol. Psychol.* 5, 403–408.
19 Horne, J.A. (1978). *Biol. Psychol.* 7, 55–102.
20 Horne, J.A. (1978). *Experientia* 33, 1175–1176.
21 Horne, J.A. (1978). *Persp. Biol. Med.* 21, 591–601.
22 Horne, J.A. (1979). *Physiol. Psychol.* 7, 115–125.
23 Horne, J.A. (1981). *Biol. Psychol.* 12, 241–290.

24 Horne, J.A. (1983). In *Sleep Mechanisms and Functions*, ed. Mayes, A., pp. 262–312. Wokingham: Van Nostrand Reinhold UK.
25 Horne, J.A. (1983). *Clin. Sci.* 65, 569–578.
26 Horne, J.A. (1985). *Ann. Clin. Res.* (in press).
27 Horne, J.A., Anderson, N.R., and Wilkinson, R.T. (1983). *Sleep* 6, 347–358.
28 Horne, J.A. and McGrath, M.J. (1984). *Biol. Psychol.* 18, 165–184.
29 Horne, J.A. and Minard, A. (1985). *Ergonomics* 28, 567–575.
30 Horne, J.A. and Moore, V.J. (1985). *Electroenceph. Clin. Neurophysiol.* 60, 33–38.
31 Horne, J.A., Percival, J.E., and Traynor, J.R. (1980). *Electroenceph. Clin. Neurophysiol.* 49, 408–413.
32 Horne, J.A. and Pettitt, A.N. (1984). *Sleep* 7, 168–169.
33 Horne, J.A. and Pettitt, A.N. (1985). *Acta Psychol.* 58, 123–139.
34 Horne, J.A. and Porter, J.M. (1975). *Nature* 256, 573–575.
35 Horne, J.A. and Reid, A.J. (1985). *Electroenceph. Clin. Neurophysiol.* 60, 154–157.
36 Horne, J.A. and Staff, L.H.E. (1983). *Sleep* 6, 36–46.
37 Horne, J.A. and Walmsley, B. (1976). *Psychophysiology* 13, 115–120.
38 Horne, J.A. and Wilkinson, S. (1984). *Psychophysiology* 22, 69–78.
39 Jones, H.S. and Oswald, I. (1968). *Electroenceph. Clin. Neurophysiol.* 24, 378–380.
40 Kales, A., Tan, T.-L., Kollar, I.J., Naitoh, P., Preston, T.A., and Malstrom, E.J. (1970). *Psychosom. Med.* 32, 189–200.
41 Karajan, I., Williams, R.L., Finley, W.W., and Hursch, C.J. (1970). *Biol. Psychiat.* 2, 391–399.
42 Kennedy, C., Kennedy, C., Gillin, J.C., Mendelson, W., Suda, S., Miyaoka, M., Ito, M., Nakamura, R.K., Storch, F.I., Pettigrew, K., Mishkin, M., and Sokoloff, L. (1982). *Nature* 297, 325–327.
43 Naitoh, P. (1976). *Waking Sleep.* 1, 53–60.
44 Palmblad, J., Petrini, B., Wasserman, J., and Akerstedt, T. (1979). *Psychosom. Med.* 41, 373–378.
45 Parker, D.C., Rossman, L.G., Kripke, D.F., Gibson, W., and Wilson, K. (1979). In *Endocrine Rhythms*, ed. Krieger, D.T., pp. 143–173. New York: Raven Press.
46 Pasnau, R.O., Naitoh, P., Stier, S., and Kollar, E.J. (1968). *Arch. Gen. Psychiat.* 18, 496–508.
47 Porter, J.M. and Horne, J.A. (1981). *Electroenceph. Clin. Neurophysiol.* 51, 426–433.
48 Quabbe, H.-J., Gregor, M., Bumke-Vogt, C., Witt, I., and Giannella-Neto, D. (1983). *Adv. Biol. Psychiat.* 11, 48–59.
49 Rechtschaffen, A., Gilliland, M.A., Bergmann, B.M., and Winterer, J.B. (1983). *Science* 221, 182–184.
50 Ruckebusch, Y. (1975). *Appl. Anim. Ethol.* 2, 3–28.
51 Sato, S., Dreifuss, F.E., and Penry, J.K. (1975). *Electroenceph. Clin. Neurophysiol.* 39, 479–489.
52 Sawka, M.N., Gonzalez, R.R., and Pandolf, K.B. (1984). *Am. J. Physiol.* 246, R72–R77.
53 Scheving, L.E. (1959). *Anat. Rec.* 135, 7–20.
54 Shapiro, C.M., Goll, C.C., Cohen, G.R., and Oswald, I. (1984). *J. Appl. Physiol.* 56, 671–677.
55 Siesjo, B.K. (1978). *Brain Energy Metabolism.* New York: Wiley.
56 Takahashi, Y. (1979). In *Functions of Sleep*, ed. Drocker-Colin, R. *et al.*, pp. 113–145. New York: Academic Press.

57 Velasco, F., Velasco, M., Cepeda, C., and Munoz, H. (1980). *Electroenceph. Clin. Neurophysiol.* **48**, 64–72.

58 Waterlow, J.C., Garlick, P.J., and Millward, D.J. (1978). *Protein Turnover in Mammalian Tissues and in the Whole Body.* Amsterdam: Elsevier.

59 Webb, W.B. and Agnew, H.W. (1970). *Science* **168**, 146–147.

60 Webb, W.B. and Agnew, H.W. (1971). *Science* **174**, 1354–1356.

61 Wilkinson, R.T. (1965). In *Physiology of Survival*, ed. Edholm, O.G. and Bacharach, A.L., pp. 399–430. London: Academic Press.

4

THE SEARCH FOR AND EVALUATION OF SLEEP-INDUCING SUBSTANCES

LIU SHIYI (S.Y. LIU),[*1] LI CHONGXI (C.X. LI),[*2]
AND XU JIEZEN (J.Z. XU)[*3]

*Shanghai Institute of Physiology, Academia Sinica, Shanghai,[*1] Department of Chemistry, Peking University, Beijing,[*2] and Shanghai Institute of Organic Chemistry, Academic Sinica, Shanghai,[*3] China*

Despite the fact that nearly one-third of our life is spent in sleep, the explanation of the mechanism of sleep remains elusive. Whether sleep is brought on by endogenous sleep-inducing substances accumulated naturally in the brain is a question that has attracted more and more attention. Though many efforts (factor S, δ-sleep inducing peptide (DSIP), sleep-promoting substance (SPS), *etc.*) have been made since 1910 (*27, 28, 30*), the mechanism is not yet known or can only be speculated. But most authors will agree that humoral control related to circadian rhythm serves as an important mechanism involved in inducing sleep and wakefulness. On the other hand, the biological relevance of sleep-inducing substances is still an unresolved question, so the survey of some quick, reliable and relatively simple methods or a set of methods for the evaluation of sleep-inducing substances seems not only interesting, but also practically necessary.

I. SEARCH FOR SLEEP-INDUCING SUBSTANCES

Since 1979 we have been interested in searching for a chemical which can induce sleep, without regard to whether it is endogenous or exogenous, sleep-inducing or-facilitating (*12, 14, 18*). We also consider that modification of chemical structure and isolation from natural products are both worthy of note. Some attention has been focused on the modification of some existing sleep-inducing or facilitating substances (benzodiazepine derivatives, DSIP, *etc.*). In addition, the possible relationship between sleep-inducing and hibernation-inducing substances has also been noted. Experience gained by our forerunners shows that the isolation of sleep-inducing or facilitating substances from natural animals or plants is quite time-consuming, but it may open broad new avenues of approach.

1. *Benzodiazepine Derivatives*
We started our research with the potent benzodiazepine derivatives which are presently the most well-known hypnotics (*4, 8*). We found

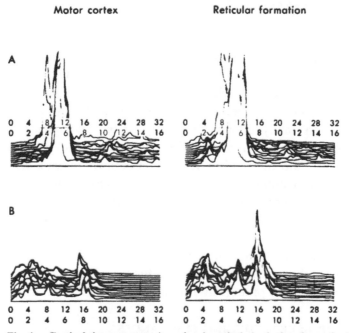

Fig. 1. Cortical (motor cortex) and subcortical (reticular formation) high frequency spindles (15–16 Hz) induced by RO-5-4200 (200 μg/kg, i.p.) in cats. A: control. B: after injection. (Band: 32 Hz, Algorithm: BERG, Filter: KORR).

that the basic feature of RO-5-4200 synthesized by us was quite different from DSIP. Besides slow wave sleep (SWS), the former could induce specific spindles with a high frequency around 16–19 c/s, which were observed predominantly in the reticular formation and motor cortex in some animals tested (Fig. 1). In the course of an extensive study of structure-activity relationships, we found that potent derivatives were seen with Cl, Br, NO_2 in the 7 position of the "A" ring and F, Cl in the 2 position of the "C" ring, but none were more potent than RO-5-4200 derivative (*32*). In the meantime, we learned from Hoffmann-La Roche that RO-20-8522 serves as an even more potent one (*29*).

2. α-Aspartyl DSIP

We noticed that DSIP is a mixture of 80% β-aspartyl DSIP and 20% α-aspartyl DSIP and that only the latter is highly active. We used tertbutyl ester to protect the carboxyl group of acidic amino acids (Asp, Glu). Using Boc to protect the α-amino group of N-terminal Trp, we used Z to protect the α-amino group of all the others, so that the Z protective group could be removed by catalytic hydrogenation under neutral conditions. Finally, the protected peptide was treated by trifluoroacetic fatty acid (TFA) to remove all protective groups simultaneously (Fig. 2). Therefore, the synthesis of Asp⁵-α-DSIP has been established. The purified sample was proved to be homogeneous *via*

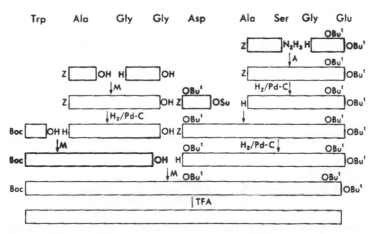

Fig. 2.   Scheme of the synthesis route of Asp⁵-α-DSIP. A=Azide method, M=Mixed anhydride method.

microcrystallinecellulose thin layer chromatography, high performance liquid chromatography (HPLC) and high voltage electrophoresis (Fig. 3) (*14*). Proton nuclear magnetic resonance (NMR) spectra indicated

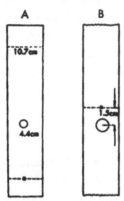

Fig. 3. A: thin layer chromatograph of microcrystallinecellulose. (Solvent system: phyridine: acetic acid: water: *n*-butanol = 12:3:10:15; detection: ninhydrin.)
B: high voltage paper electrophoresis. (Buffer: pyridine: acetic acid: water = 1:10:89; voltage: 1900 V, electric current 35–42 mA., detection: ninhydrin; paper length: 52 cm.)

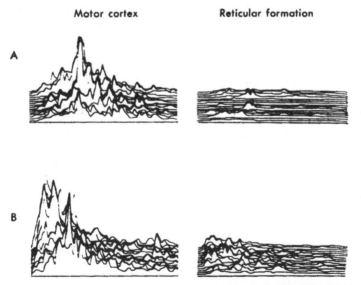

Fig. 4. Cortical (motor cortex) and subcortical (reticular formation) SWS EEGs induced by Asp^6-α-DSIP (50 μg/kg, i.p.) in cats. A: control. B: after injection. (Band: 16 Hz, Algorithm: FOU, Filter: KORR).

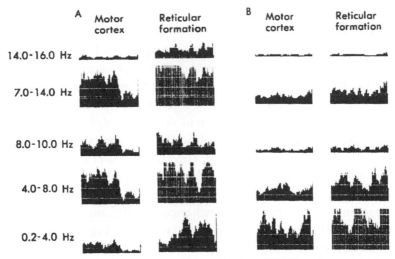

Fig. 5.   Trends computation of the effects of Asp⁵-α-DSIP (50 μg/kg, i.p.) in free-moving cats. A: control, B: after Asp⁵-α-DSIP. (Gain: 0.5, TC: 0.03, FOU, Band selectable within 0–16 Hz).

that the methylene proton could be seen only for the Asp⁵-α-DSIP at $\delta = +1.96$ ppm. Asp⁵-α-DSIP is known to induce SWS in rabbits as well as in cat (Figs. 4 and 5). It has also been known that there is no obvious sign of adaptation to this peptide, which failed to show the "dose-effect-curve" observable for usual drugs (*17*). The basic identity between Asp⁵-α-DSIP synthesized by tetrahydrothiazole-2-thione (TTT) and those prepared by other methods could be established.

3.   Phe⁵-series
After synthesis of a few dozen analogues we first reported a new analogue (Trp-Ala-Gly-Gly-Phe-Ala-Ser-Gly-Glu), which was also found to be active (*13*). Attention has been focused on the repeated synthesis of Phe⁵-DSIP and Phe⁵-series by the solid phase as well as the liquid phase method. It is interesting to note that among the known amino acids only Trp and Phe *per se* could induce delta-enhancing effects. From the theoretical and practical point of view, short Phe⁵-series (*e.g.*, MW <600) will be what interests us most. It is hoped that further modification of chemical structure will make the finding of a sleep-peptide with smaller molecular weight or more specificity possible.

4. Oxoquinoline Derivative

Recently, we isolated some crystalline compounds from the fruit extract of *Parthenocissus tricuspidata* (Sieb et Zucc) planch. One of them is an oxoquinoline derivative which seemed to have some hypnotic effects in mice (*34*). L-Trp, a precursor of 5-hydroxytryptamine (5-HT), is considered as an important amino acid related to sleep, but some controversy exists concerning the role of 5-HT. We observed that after administration of Asp⁵-α-DSIP, some elevation of 5-HT contents could be seen in the thalamus and hypothalamus in rats (*33*), but we also found that during stepping-inducing substances (SIS) induced stepping 5-HT contents were significantly enhanced in the midbrain in guinea pigs (*19*). Others also reported that in freely moving cats increases of serotonin-containing midbrain Raphe unit activity caused movements rather than sleep. It is known that after substitution of the indole ring in Trp by a napthalene ring, the modified analogue could also produce hypnotic effect with no change of 5-HT in the brain (*10*).

5. Tupaiidae

With regard to mammals, attention has been focused on the Tupaiidae, although pigs have interested us as well. Tupaiidae—the "living model" of the oldest primates—are phyletically closer to the primates than goats and rodents. Tupaiidae have only two subfamilies and most of them are distributed in South Asia. Tree shrews (*Tupaia belangeri chinensis*) showed striking differences in activities between morning and night (*21*). They bustle in continuous jerky movements in the morning, but are also apt to sleep. Their total sleep time in terms of hours per day is 15–16 *versus* 2–4 in goats. Our preliminary amino acid analyses of urine in tree shrews indicate that enhancement of some amino acids (*e.g.*, Gly, Glu) could be seen during sleep (*22*). It seems reasonable that different animals share analogous, if not identical sleep-inducing substances in the brain. This question interests us, but it can only be answered by further research.

II. EVALUATION OF SLEEP-INDUCING SUBSTANCES

The assay of sleep-inducing substances seems more difficult and elusive than that of motor-activity (*e.g.*, stepping automatism or stereotyped movements)-inducing substances. Long stretches of electroencephalogram (EEG) profiles and their short term computer analysis are the commonly used method to evaluate the former (*39*), whereas the latter can be easily and convincingly demonstrated almost at a glance (*16*). Sleep EEGs contain more stationary or quasi-stationary segments with few transients, so the survey of some quick methods of compressed evaluation instead of time-consuming inspection of long stretches of unchanging (*i.e.*, stationary or quasi-stationary) records seems useful. On the other hand, conventional methods have been predominantly applied to analysis of the patterns of cortical spontaneous activity, so little information is available for the dynamic evaluation of both cortical and subcortical sleep EEGs or the responsiveness of spontaneous activity. It is also of interest to see whether sleep-inducing substances can be assayed by the long term circadian rhythm method, as well as the ultrashort term method *in vitro*.

1. Compressed EEG Evaluation

An attempt has been made to evaluate sleep-inducing (Asp^5-α-DSIP) and sleep-facilitating (RO-5-4200) substances by using relative long term power spectra array or trends computation (Schwarzer), automatic adaptive segmentation and clustering (*1, 23*), as well as "EEG-EPograms" (*2, 3*), respectively. Power spectra array or trends computation based on the Fourier or Berger (BERG) (*31*) transform provides a quick method of relative long term (up to 130,560 sec) evaluation which can be viewed at a glance. Figures 1, 4, and 5 present examples of conventionally used methods. Data reduction of overwhelming sleep EEGs based on automatic adaptive segmentation and clustering also seems promising. It is of interest whether a 2D or 3D representation will improve the data reduction to 100 : 1 or more and compress the overall data such that it can be visually viewed almost at a glance. The EEGs are viewed through a moving window, which is compared with the earlier fixed reference window. The method of segmentation is based on the autocorrelation function (ACF). If the difference between EEG pieces seen through these two windows is large enough, then a new segment line is recognized. The difference measure through these two windows is estimated from the combination of a

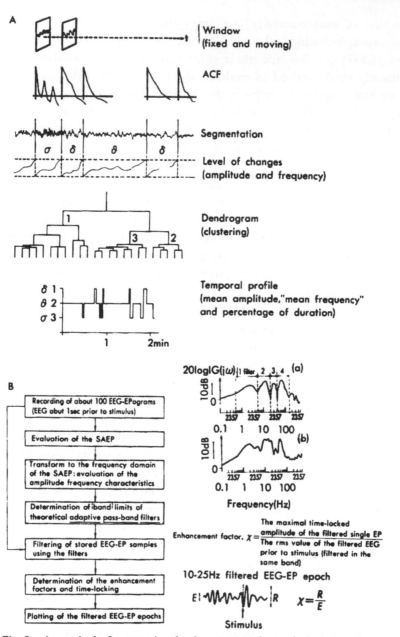

Fig. 6. A: method of automatic adaptive segmentation and clustering. B: method of "EEG-EPograms." EP, evoked potential; SAEP, selectively averaged EP.

percent difference measure for amplitude and frequency, respectively. The two ACFs are normalized and compared. The segments are then clustered by means of the cluster algorithm (1, 35). The clustered segments and dendrograms are written out, together with a temporal profile (Fig. 6A). Peak amplitude, peak frequency, and percentage of duration of the overall principal clusters are then illustrated in a 2D representation. An attempt has also been made to compute the peak frequency and amplitude of each principal cluster with the power spectra based on the autocorrelation functions and algebraic family of data windows (23). "EEG-EPograms" contains only a portion of EEG just prior to a stimulus (a sound or a light) and an evoked potential (EP) just following the stimulus (1 sec–1 sec). This method allows us not only to evaluate the spontaneous activity, but also the responsiveness and excitability of brain potentials or the reorganization of spontaneous activity. It is different from the conventional EP research, which consists of the application of averaging techniques and peak analysis. The EP are averaged using a selective averaging method. The selectively averaged EP (SAEP) are transformed to the frequency domain with Fourier transform in order to obtain the amplitude frequency characteristic, $|G\ (jw)|$, of the cortical or subcortical regions studied. EEG-EP sets are filtered with response adaptive filters. The voltage of the root-mean-square (rms) values of maximal amplitude existing in the filtered EPs and the so-called enhancement factor for the given EEG-EPograms are evaluated (Fig. 6B) (3). The histogram of the poststimulus amplitude frequency characteristics show the frequency stabilization, time-locking and amplification of spontaneous activity upon stimulation. Stimulation brings the brain into a more coherent, sharply defined state. By pulling various oscillators into synchrony, the brain performs an instant transition from a chaotic regime to a more ordered regime. It is worthy of note that the "EEG-EPograms" may serve as a method for the dynamic evaluation of sleep-inducing substances. The combined application of the above mentioned compressed EEG evaluation showed the exceptional similarity of Asp⁵-α-DSIP and the dissimilarity of RO-5-4200 to natural sleep. Both could induce SWS EEGs in cortical and subcortical regions studied in rabbits, but the specific spindle with a high frequency around 15–16 Hz could be observed only by RO-5-4200. Moreover, the former (50 μg/kg, i.p.) could induce cortical and subcortical SWS EEGs in three of the five cats in free-moving conditions (Fig. 5), whereas the latter (200 μg/kg, i.p.) induced excitation instead of sleep in concomitance with significant specific spindles in all four cats studied (*24*).

2. Circadian Rhythm Evaluation

We used tree shrews (Tupaiidae) for our circadian rhythm evaluation. Effects of DSIP (same dose of 60 μg/kg for six consecutive days) on 24-hr motor activity were studied in 22 tree shrews (*T. belangeri chinensis*) weighing 100–146 g. They were exposed to artificial L12 (6:00–18:00) and D12 (18:00–6:00) schedules. DSIP and 0.9% NaCl solution were administered (i.p.) at 6:00 (morning group) and 18:00 (evening group). Enhancement of the 24-hr motor activity by DSIP could be seen only in the latter group. Moreover, this enhancement was mainly concentrated at 06:00–10:00 and 14:00–18:00, which serve as the periods of maximal motor activity of tree shrews in a natural environment (*38*). The above results suggest that circadian rhythm may serve as a recognizable method for evaluation, but further efforts seem necessary to elucidate whether this feature is specifically available only for sleep-inducing substances.

3. In Vitro Evaluation

Some attention has also been focused on the possibility of evaluating sleep-inducing substances *in vitro*. The widely adopted method of guinea-pig ileum (GPI), mouse vas deferens (MVD), and rat vas deferens RVD) for the identification of μ, δ, and ε receptors has been shown to be a quick method for the evaluation of opioid peptides (*25*), but no information is available for sleep-inducing substances. Since results indicate that Asp^5-α-DSIP (0.1–100 μM) exerted no appreciable changes on GPI, MVD, and RVD tested (*20*), attention has been transferred to the thin cortical or hippocampal slice, as well as to the intact brain-spinal cord slice *in vitro*. Sleep-inducing substances may exert some effect on extracellular field potentials and on intracellular excitatory and inhibitory post-synaptic potentials *in vitro*. 400–500 μm hippocampal slices of guinea pig were placed in the tissue chamber, the bathing medium of which was continuously areated by 95/5% O_2/CO_2. Stimulating electrodes were placed in Schaffer's collaterals and in the alveus, which allow orthodromic or antidromic activation, respectively. Field

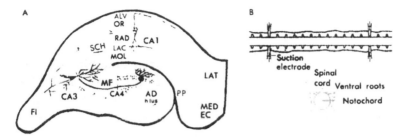

Fig. 7. A: the hippocampal slice preparation *in vitro*. B: the lamprey spinal cord preparation *in vitro*. (Preparation was bathed in cooled, oxygenated lamprey physiological fluid (8–9 C). Cross-section shows the exposed spinal cord with its ventral roots on top of the notochord).

potentials were recorded in the apical or basal dendritic, or in the somatic region of CA1 pyramidal cells (Fig. 7A) (*5, 15*). Micromolar concentrations of RO-5-4200, and to a lesser extent of Asp^5-α-DSIP, could attenuate neurotransmission resulting from the orthodromic activation *in vitro*. On the other hand, at least three types of excitatory amino acid receptors have been described: 1) N-methyl-D-aspartate (NMDA) activated and 2-amino-5-phosphonovalevate (2APV) or D-α-aminoadipate (DαAA) antagonized; 2) Quisqualate activated and L-glutamic acid diethyl ester (GDEE) antagonized; and 3) kainate activated with no known specific antagonist (*7, 26*). In the meantime, there is considerable interest in the function of excitatory amino acid in the vertebrate or amphibian spinal cords. Ventral activities have been known to be evoked by activation of NMDA-receptors in lamprey spinal cord *in vitro* (*11*). The spinal - notochord preparation of adult *Ichthyomyzon unicuspis* or *Lampetra fluviatillis* (*12*) was dissected free from muscles into 15–30 segments. The preparation was placed in an *in vitro* chamber and pinned down in "Sylgard" filled with cooled (7–9°C) lamprey physiological solution equilibrated with 95/5% O_2/CO_2 (*36*). Ventral root activities were recorded *en passant* close to the lateral edge of the spinal cord with glass suction electrodes (Fig. 7B). Surprisingly, RO-5-4200 (1–100 μM) could facilitate N-methyl-DL-aspartate, (NMA)-induced ventral root activities in lamprey spinal cord *in vitro*. Moreover, a narrow dose of Asp^5-α-DSIP (around 1 μM) seemed to have a similar effect, but to a lesser extent.

SUMMARY

We started our research from the potent benzodiazepine derivatives. In addition, the synthesis of Asp⁵-α-DSIP has been established. There is no obvious sign of adaptation to this peptide, which failed to show the "dose-effect-curve" commonly observed for drugs. After synthesis of a few dozen analogs we reported a new analog (Trp-Ala-Gly-Gly-Phe-Ala-Ser-Gly-Glu), which was found to be active also. Isolation of sleep substances from natural plants and animals has been shown to be time-consuming, but has the potential to open broad new areas of research.

The assay of sleep-inducing substances seems more difficult and elusive than that of the motor-activity-inducing substances. The survey of some quick methods of compressed EEG evaluation (*e.g.*, relative long term power spectra array or trends computation, automatic adaptive segmentation and clustering, "EEG-EPograms", *etc.*) seems useful. It is also of interest whether the sleep-inducing substances can also be assayed by the long term circadian rhythm method, as well as the ultra-short term method *in vitro* (*e.g.*, thin or intact slice). It is hoped that the combined application of both *in vivo* and *in vitro* methods will be favorable for the dynamic evaluation of sleep-inducing substances.

REFERENCES

1 Barlow, J.S., Creutzfeldt, O.D., Michael, D., Houchin, J., and Epelbaum, H. (1981). *Electroenceph. Clin. Neurophysiol.* 51, 512–525.

2 Basar, E., Demir, N., Gönder, A., and Ungan, P. (1979). *Biol. Cybernet.* 34, 1–19.

3 Basar, E. (1980). *EEG-Brain Dynamics: Relation between EEG and Brain Evoked Potentials.* Amsterdam: Elsevier.

4 Borbély, A.A., Mattmann, P., Leopfe, M., Fellmann, I., Gerne, M., Strauch, I., and Lehmann, D. (1983). *Eur. J. Pharmacol.* 89, 157–161.

5 Buckle, P.J. and Haas, H.L. (1982). *J. Physiol.* 326, 109–122.

6 Haines, D.E., Murray, H.M., Albright, B.C., and Goode, G.E. (1974). In *Perspectives in Primate Biology*, ed. Chiarelli, A.B., pp. 29–92. New York: Plenum Press.

7 Davies, J., Francis, A.A., Jones, A.W., and Watkins, J.C. (1981). *Neurosci. Lett.* 21, 77–81.

8 Depoortere, H. and Granger, P. (1982). In *Sleep 1982*, ed. Koella, W.P., pp. 286–287. Basel: Karger.

9 Durrani, T.S. and Nightingale, J.M. (1972). *Proc. IEE* 119, 343–352.

10 Fornal, C., Wojeik, W.J., Radulovacki, M., and Schlossberger, H.O. (1979). *Pharmacol. Biochem. Behav.* 11, 319–323.

11 Grillner, S., McClellan, A., Sigvardt, K., Wallen, P., and Wilen, M. (1981). *Acta Physiol. Scand.* **113**, 549–551.

12 Hardisty, M.W. and Potter, I.C. (1971). *The Biology of Lampreys*, vol. 1. London: Academic Press.

13 Hsu, J.Z., Cheng, L.L., Wang, S.I., Shen, W.Z., Chien, C.H., Huang, J.X., Liu, S.Y., Chang, W.Y., Tai, S.C., and Wang, J.S. (1980). In *Nucleic Acid and Proteins*, ed. Shen, Z.W., pp. 206–211. Beijing: Science Press.

14 Ji, A.X., Li, C.X., Ye, Y.H., Lin, Y., Xin, Q.Y., Liu, S.Y., Zhong, W.Y., Wang, Z.S., and Dai, X.J. (1983). *Sci. Sinica B*, **226**, 174–185.

15 Kuhnt, U., Kelly, M.J., and Schaumberg, R. (1979). *Exp. Brain Res.* **35**, 371–385.

16 Liu, S.Y., Zhang, W.Y., Tai, X.J., and Wang, Z.S. (1980). *Kexue Tongbao* **25**, 562–565.

17 Liu, S.Y., Zhang, W.Y., Wang, Z.S., and Tai, X.J. (1981). *Acta Physiol. Sinica* **34** 90–97.

18 Liu, S.Y. (1982). *Acta Psychol. Sinica* **14**, 19–28.

19 Liu, S.Y., Guang, P.G., Xu, B., and Bai, F.L. (1983). *Neurosci. Lett.* (Suppl. 14) S341.

20 Liu, S.Y. and Wang, H.Y. Unpublished data.

21 Liu, S.Y., Zhang, W.Y., Wang, Z.S., Dai, X.J., and Li, C.D. (1982). In *Sleep 1982*, ed. Koella, W.P., pp. 226–228. Basel: Karger.

22 Liu, S.Y., Li, C.X., Hsu, J.Z., and Cheng, D.M. (1983). 4th International Sleep Congress, p. 117.

23 Liu, S.Y., Michael, D., and Creutzfeldt, O.D. Unpublished data.

24 Liu, S.Y., Basar, C., Rosen, B., and Basar, E. Unpublished data.

25 Martin, W.R., Endes, C.G., Thompson, J.A., Huppler, R.E., and Gilbert, P.E. (1976). *J. Pharmacol. Exp. Ther.* **197**, 517–532.

26 McLennan, H., Hicks, T.P., and Hall, J.G. (1981). In *Amino Acid Neurotransmitters*, ed. De Feudis, F.V. and Mandel, P., pp. 213–221. New York: Raven Press.

27 Nagasaki, H., Iriki, M., Inoué, S., and Uchizono, K. (1974). *Proc. Japan Acad.* **50**, 241–246.

28 Pappenheimer, L.R., Koski, G., Fencl, V., Karnovsky, M.L., and Krueger, J. (1975). *J. Neurophysiol.* **38**, 1299–1311.

29 Scherschlicht, R. and Marias, J. (1982). Report of Hoffmann-La Roche Co. Ltd. No. B-89'363, pp. 1–20.

30 Schoenenberger, G.A., Maier, P.F., Tobler, H.J., Wilson, K., and Monnier, M. (1978). *Pflügers Arch.* **376**, 119–129.

31 Sciarretta, G. and Ereuliani, P. (1975). Symposium of the Study Group for EEG-Methodology, pp. 487–496.

32 Tang, Z.J., Zu, S.S., Zhang, S.A., Jiu, S.Y., and Zhang, W.Y. Unpublished data.

33 Wang, H.Y., Wang, Z.S., Zhang, W.Y., and Liu, S.Y. (1984). *Abstr. 3rd Symp. Physiopsychol.*, All-China Society for Psychological Sciences, Beijing, p. 71.

34 Wang, Y.F., Yao, R.R., Zhou, W.S., Liu, S.Y., Zhang, W.Y., and Wang, Z.S. Unpublished data.

35 Ward, J.H. (1963). *J. Am. Stat. Assoc.* **58**, 236–244.

36 Wickelgren, W.O. (1977). *J. Physiol.* **270**, 89–114.

37 Chi, A.H., Li, C.H., Yieh, Y.H., Lin, Y., Lu, Y.J., Hsing, C.Y., Liu, S.Y., Chang, W.Y., Wang, T.S., and Tsi, S.C. (1981). *7th American Peptide Symposium*, pp. 53–56. Rockford: Pierce.

38 Zhang, W.Y., Wang, Z.S., Wang, H.Y., Liu, S.Y., and Schoenenberger, G.A. (1984).

Abstr. 3rd. Symp. Physiopsychol., All-China Society for Psychological Sciences, Beijing, pp. 71–72.

39 Ziegler, W.H. and Schalch, E. (1982). In *Sleep 1982*, ed. Koella, W.P., pp. 427–429. Basel: Karger.

EVOLUTIONARY AND
ADAPTIVE FEATURES

5

DEPRIVATION OF SLEEP AND REST IN VERTEBRATES AND INVERTEBRATES

IRENE TOBLER

Institute of Pharmacology, University of Zürich, CH-8006 Zürich, Switzerland

I. SLEEP DEPRIVATION IN MAMMALS

Sleep deprivation (SD) experiments in man and rat have revealed an important characteristic of sleep. Sleep is apparently regulated relative to an internal reference level. Prolonged waking leads to a compensatory increase of sleep (see Borbély, Chapter 2) whereas excess sleep results in a decrease of slow wave sleep (5). The electroencephalogram (EEG) has been the main parameter by which the effects of SD have been characterized. Prolonged SD elicited an increase of slow wave activity as well as a REM sleep rebound in rat and man (1, 3, 11). It is important to note that paradoxically SD may also have an activating effect (see Borbély, Chapter 2).

A compensatory increase in EEG parameters after prolonged waking has been described for several mammalian species (Table I). The data are consistent with the hypothesis that the close relationship between sleep and prior waking is a general feature of mammalian sleep. Analysis of the mechanisms involved in sleep regulation may provide insight into the functional significance of sleep.

TABLE I

Species Which Have Been Subjected to Total Sleep (Rest) Deprivation and Whose EEG or Motor Activity was Investigated during Recovery

Species	SD (hr)	EEG slow wave sleep δ activity	Motor activity	Others	References
Mammals					
Rhesus monkey	88	+			21
Macaque mulatta	176	+			21
Dog	12	+			27
	3, 6, 12	+		Hormones	28
	24		−		Tobler and Sigg (in preparation)
Cat	7 days	+			15
	72	PS: +			13
	12, 24	+		Sleep latency	30
Bottlenose dolphin	60	+			16
Donkey	48	+(delayed)		Aggressiveness	23
Rabbit	24, 28	+		Awakening	20
Rat	216			Exhaustion	31
	12	+		Awakening thresholds	10
	24	+			11
	12,24	+	−		1
Birds					
Barbary dove	6, 36			Eye closure	14
Reptiles					
Caiman	24, 48			Spikes and	9
Iguanid lizard	48			"Sleep"	6
Box turtle	48			Latency	8
Red footed tortoise	48				7
Sea turtle	12	No change			26
Fish					
Perch	8, 12	+ and −			Tobler and Borbély (in press)
Guppie	96		−	"Rest" Latency	24
Invertebrates					
Cockroach	3		−		29

Slow wave sleep or δ sleep +: enhancement; motor activity −: decline; SD, duration of deprivation; PS, paradoxical sleep.

II. MOTOR ACTIVITY

In several studies involving SD motor activity was also measured. Borbély and Neuhaus (*1*) subjected rats to 24 hr SD and continuously recorded the vigilance states and motor activity. When recovery coincided with the dark period, the habitual active circadian phase of the rat, total sleep in the first 12 recovery hours was significantly enhanced while motor activity was reduced to 84.16±8.93% SEM (n.s.) of baseline (see also Fig. 1). Similarly, in man Naitoh *et al.* (*18*) demonstrated a reduction of body movements during recovery from SD. In

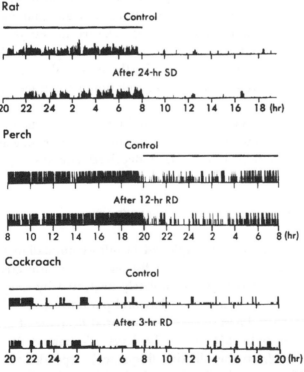

Fig. 1. Motor activity of an individual rat, perch, and cockroach measured continuously for 24 hr during a control day and during a recovery day immediately following 24-hr SD, or 8 hr and 3 hr rest deprivation (RD). Deprivation ended with dark onset (rat, cockroach) or light onset (perch). Black bar indicates the 12-hr dark period of the 12 hr L : 12 hr D cycle. Activity bars are explained in legend of Fig. 4.

another study with human subjects, total values of short episodes with movements (Types A, B, C, and D, classified according to Spiegel (*25*)) were reduced after 40.5 hr of SD (77.2% of baseline, 100% = mean value of 2 baseline nights; $p<0.025$; Eglin and Borbély, unpublished data).

To further validate the use of measures of motor activity I subjected laboratory dogs to sleep deprivation. Motor activity was continuously recorded with a portable activity monitor worn around the neck. Motor activity measures were obtained under normal conditions and after 24 hr SD. Recovery was allowed during the light period, the habitual active phase of the laboratory dogs. Motor activity values expressed as number of episodes with suprathreshold values were reduced during the light period (30.04% of control; $p<0.02$) as well as during the following 12 hr dark period (85.46%; $p<0.1$; Tobler and Sigg, in preparation). The experiments mentioned above indicate that motor activity may be a simple and easily obtained measure to test the hypothesis that the close relationship between sleep and prior waking is a general feature of mammalian sleep. Furthermore, SD studies in other vertebrate classes may reveal if sleep regulation is a unique property of mammalian sleep.

The physiological mechanisms involved in sleep regulation may be related to those responsible for sleep changes induced by endogenous substances that were obtained from sleep deprived animals (*e.g.*, *2*, *17*, *19*).

III. SLEEP DEPRIVATION IN OTHER VERTEBRATES

The method of SD has been little exploited in infra-mammalian sleep studies. Few species of birds, reptiles, and fish have been subjected to SD, whereas amphibians have been completely neglected by sleep researchers (Table I). The measure of a behavioral parameter, eye blinking frequency, in barbary doves showed a substantial decrease of blinking frequency after 6–36 hr SD (*14*). The level of vigilance during recovery (as estimated from the blinking frequency) depended on the length of prior SD. Sleep deprivation for 48 hr in five species of reptiles, caiman, box turtle, red-footed tortoise and two iguanid lizards elicited a large increase in EEG spike activity during recovery (*6–9*). Twelve hours SD in one specimen of sea turtle had no effect on motor activity in the subsequent 8 hr recording period (26).

The measure of behavioral states of a schooling fish population revealed no resting behavior during light exposure for 96 hr. The fish were therefore deprived of rest by this manipulation. During recovery from this schedule latency to sleep behavior was shortened (24). Our own studies with perch involved the following two activating procedures: 1) mechanical stimulation and 2) activation by 6 and 12 hr light during the habitual dark phase. Motor activity was continuously measured by time-lapse video recording. Mechanical stimulation was followed by a significant enhancement of activity, while stimulation by 12-hr light induced a significant decrease of activity during recovery (Fig. 1). In these fish it became apparent that certain deprivation procedures may induce activation which possibly interferes with the expression of a concomitant rest compensation. Such experiments demonstrate the importance of applying an adequate sleep deprivation procedure with a minimum of stress and other unspecific factors. Sleep deprivation or rest deprivation experiments with species belonging to infra-mammalian classes indicate that also these animals may exhibit compensatory effects after prolonged waking. This implies that the homeostatic regulation of rest and activity is not restricted to mammals. The presence of humoral rest-promoting factors in non-mammalian species has not yet been unambiguously demonstrated.

IV. SLEEP DEFINITION

The investigation of sleep and its function is strongly influenced by the traditional criteria used for the definition of sleep. Due to the general application of electrographic criteria, sleep research is presently centered on vertebrate species. The study of sleep in vertebrates involves, however, a very selective part of the animal kingdom, since it represents only 3.9% of all known species (Fig. 2). Because of the close relationship between EEG and behavior in virtually all mammals and birds thus far studied, little effort has been made to systematically quantify behavioral manifestations of sleep. This limitation hampers the study of the evolutionary aspects of sleep. It is therefore essential that definitions of sleep

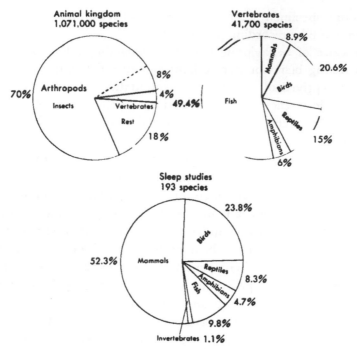

Fig. 2. Species representation in the animal kingdom (above) *versus* representation in studies investigating the presence of sleep or sleep-like states. Anecdotal reports were not included. For references see Campbell and Tobler (*4*).

in infra-mammalian vertebrates as well as in invertebrates be established on the basis of behavioral manifestations. Possibly electrographic indices may be found to correlate with behavioral states. A unique example of this approach is the work of Kaiser and Steiner-Kaiser (*12*) who demonstrated in honey bees an association between electrographic changes in neurons and alterations in behavioral response.

I have proposed that sleep should be defined not only *descriptively* (*i.e.*, by traditional electrophysiological criteria and by behavioral criteria as proposed by Piéron (*22*) and extended by Flanigan (*6*)), but also *functionally* in terms of its regulatory property (*29*). If the concept of sleep is broadened in this way, a wide range of animals may be found to exhibit a sleep-like state and therefore become accessible for sleep research.

V.   THE REST-ACTIVITY RHYTHM

The circadian rest-activity rhythm may constitute a phylogenetic precursor of sleep. Most animals that have been studied exhibit a rest-activity rhythm. Rest and activity are mutually exclusive states. While *activity* can be clearly defined by the behavior of the animal, *rest* is defined merely by the absence of activity. It is, however, an inhomogenous state. In mammals the sleep-wake rhythm does not necessarily coincide with the rest-activity rhythm (Fig. 3). By combining EEG and motor activity measures, sleep researchers can easily discern if an animal is awake and resting or if it is asleep. Those animals which are not considered to sleep by these criteria may serve to identify a sleep-like sub-state of rest. However, systematic and quantitative analysis of resting behavior has rarely been undertaken. Kaiser and Steiner-Kaiser (*12*) employing electrophysiology and behavioral criteria in the honey bee revealed a state that is comparable to sleep in vertebrates. Presently we are investigating resting behavior in other arthropods. Continuous measures of rest and activity are being obtained by time-lapse video recording. Compared to the cockroach, scorpions show little and irregular activity (Fig. 4). A closer analysis of rest in these animals allowed a classification of consecutive resting behaviors characterized by

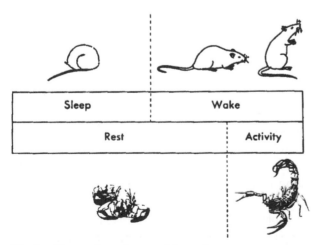

Fig. 3.   Sleep-wake and rest-activity cycle.

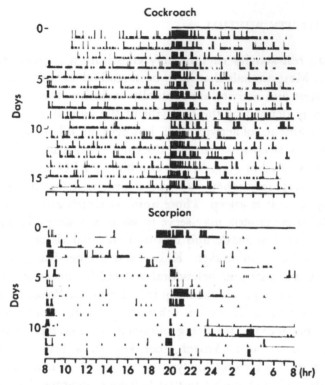

Fig. 4. Activity plots of an individual cockroach (1 bin=2.5 min) or scorpion (1 bin= 1.66 min) recorded continuously by time-lapse video recording for 16 d and 13 d respectively. High bars: locomotor activity; low bars: no locomotion, with limb or antenna movement (cockroach) or limb or tail movement (scorpion); intervals between bars: inactivity. Dark period indicated by horizontal bar. Abscissa: time of day.

body, limb, and tail position. Current experiments will reveal if selective deprivation of resting states is compensated. I have shown that enforced activity during part of the habitual rest period in an insect elicited a reduction of activity during the recovery period (Fig. 1; (29)). The results indicated that rest or immobility in an animal for which sleep cannot be defined by electrophysiological criteria may in part depend on prior waking time. Therefore, at least in one arthropod species, rest is not only a function of the circadian rest-activity rhythm, but is determined also by other regulatory mechanisms. In this respect rest in cockroaches is comparable to sleep in vertebrates. Further studies focussing on the resting state are required to investigate whether an intensity parameter comparable to slow waves in mammals can be discriminated also in invertebrates.

The study of rest and sleep-like states and of their evolution in the animal kingdom may help to understand the mechanisms and functions of sleep.

SUMMARY

Sleep deprivation in mammals elicits a compensatory response which consists of enhanced slow wave sleep or EEG slow wave activity, REM sleep rebound and reduction of motor activity. This regulatory property of sleep should be examined in animals which, due to the absence of traditional EEG criteria, are considered not to sleep. Since rest-activity rhythms are ubiquitous, investigations of rest, a state from which sleep may have evolved, could reveal elementary properties of sleep. Rest deprivation experiments in infra-mammalian vertebrates and in invertebrates may contribute to trace the phylogenetic origins of sleep. Sleep and rest deprivation in such diverse species as rat, dog, fish and cockroach revealed compensatory responses during the subsequent recovery period. Recordings were restricted to motor activity, a simple and easily measurable parameter. The results indicate that both sleep and rest are not exclusive functions of the circadian rest-activity rhythm, but are determined by additional regulatory mechanisms.

Acknowledgment

The studies were supported by Swiss National Science Foundation Grant 3.518–0.83.

REFERENCES

1 Borbély, A.A. and Neuhaus, H.U. (1979). *J. Comp. Physiol.* **133**, 71–87.
2 Borbély, A.A. and Tobler, I. (1980). *Trends Pharm. Sci.* **1**, 356–358.
3 Borbély, A.A., Baumann, F., Brandeis, D., Strauch, I., and Lehmann, D. (1981). *Electroenceph. Clin. Neurophysiol.* **51**, 483–493.
4 Campbell, S.S. and Tobler, I. (1984). *Neurosci. Biobehav. Rev.* **8**(3), 269–300.
5 Feinberg, I., Fein, G., and Floyd, T.C. (1980). *Electroenceph. Clin. Neurophysiol.* **50**, 467–476.
6 Flanigan, W.F., Jr. (1973). *Brain Behav. Evol.* **8**, 401–436.
7 Flanigan, W. (1974). *Arch. Ital. Biol.* **112**(3), 253–257.

8 Flanigan, W.F., Knight, C.P., Hartse, K.M., and Rechtschaffen, A. (1974). *Arch. Ital. Biol.* **112**, 227–252.

9 Flanigan, W.F. Jr., Wilcox, R.H., and Rechtschaffen, A. (1973). *Electroenceph. Clin. Neurophysiol.* **34**, 521–538.

10 Frederickson, C.J. and Rechtschaffen, A. (1978). *Sleep* **1**, 69–82.

11 Friedman, L., Bergmann, B.M., and Rechtschaffen, A. (1979). *Sleep* **1**, 369–391.

12 Kaiser, W. and Steiner-Kaiser, J. (1983). *Nature* **301**, 707–709.

13 Kiyono, S., Kawamoto, T., Sakakura, H., and Iwama, K. (1965). *Electroenceph. Clin. Neurophysiol.* **19**, 34–40.

14 Lendrem, D.W. (1982). In *Sleep 1982*, ed. Koella, W.P., pp. 134–138. Basel:Karger.

15 Lucas, E.A. (1975). *Exp. Neurol.* **49**, 554–568.

16 Mukhametov, L.M., Supin, A.Y., and Polyakova, I.G. (1977). *Brain Res.* **134**, 581–584.

17 Nagasaki, H., Iriki, M., Inoué, S., and Uchizono, K. (1974). *Proc. Japan Acad.* **50**, 241–246.

18 Naitoh, P., Muzet, A., Johnson, L.C., and Moses, J. (1973). *Psychophysiology* **10**, 363–368.

19 Pappenheimer, J.R. (1983). *J. Physiol.* **336**, 1–11.

20 Pappenheimer, J.R., Koski, G., Fencl, V., Karnovsky, M.L., and Krueger, J. (1975). *J. Neurophysiol.* **38**, 1299–1311.

21 Pegram, G.V., Reite, M.L., Stephens, L.M., Crowley, T.J., Buxton, D.F., Lewis, O.F., and Rhodes, J.M. (1969). 6571 Aeromedical Res. Laborat., Holloman, AFM, N.M. Report ARL-TR-69-19, pp. 47–53.

22 Piéron, H. (1913). *Le Problème Physiologique du Sommeil*. Paris: Masson.

23 Ruckebusch, Y. (1970). *Cah. Méd. Vét.* **39**, 210–225.

24 Shapiro, C.M. and Hepburn, H.R. (1976). *Physiol. Behav.* **16**, 613–615.

25 Spiegel, R. (1981). *Sleep and Sleeplessness in Advanced Age*. MTP Press Limited, Falcon House.

26 Susic, V. (1972). *J. Exp. Marine Biol. Ecol.* **10**, 81–87.

27 Takahashi, Y., Ebihara, S., Nakamura, Y., and Takahashi, K. (1978). *Neurosci. Lett.* **10**, 329–334.

28 Takahashi, Y., Ebihara, S., Nakamura, Y., and Takahashi, K. (1981). *Endocrinology* **109**, 262–272.

29 Tobler, I. (1983). *Behav. Brain Res.* **8**, 351–360.

30 Ursin, R. (1971). *Acta Physiol. Scand.* **83**, 352–361.

31 Webb, W.B. and Agnew, H.W. (1962). *Science* **136**, 1122.

6

UNIHEMISPHERIC SLOW WAVE SLEEP IN THE BRAIN OF DOLPHINS AND SEALS

LEV M. MUKHAMETOV

Institute of Evolutionary Morphology and Ecology of Animals, USSR Academy of Sciences, Moscow, 117071, U.S.S.R.

Investigating the sleep in marine mammals, we have found a new form of sleep which we called a unihemispheric slow wave sleep. At present, unihemispheric slow wave sleep has been discovered in two species of dolphins, namely, in the bottlenose dolphin (*Tursiops truncatus*) (*8, 10*) and in the porpoise (*Phocoena phocoena*) (*7*), as well as in one species of pinnipeds, in the northern fur seal (*Callorhinus ursinus*) (*6*). No unilateral electroencephalogram (EEG) slow waves were found in the two other species of Pinnipedia which we studied, the Caspian seal (*Phoca caspica*) (*9*) and the harp seal (*Pagophilus groenlandicus*). Other publications on electrophysiological sleep research in pinnipeds (*1, 12*) do not contain any information on the interhemispheric interrelations during sleep. No unihemispheric slow waves have been found in our studies of the Caribbean manatee (*Trichechus manatus*) (*13*), but so far this conclusion is only preliminary, because we have examined only one animal during two nights. There are grounds to hope that further research will reveal unihemispheric sleep in other marine mammals from Cetacea, Sirenia, and Pinnipedia.

Unihemispheric slow wave sleep is defined as a state of the brain in

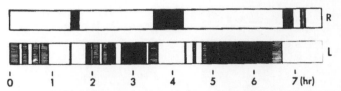

Fig. 1. Diagrams of EEG stages in the brain hemispheres of a bottlenose dolphin during a night. ☐ desynchronization; ▤ intermediate synchronization; ■ delta synchronization. R, right hemisphere; L, left hemisphere. Bipolar recordings from roughly symmetrical areas of the parietal cortex. Time scale in hours.

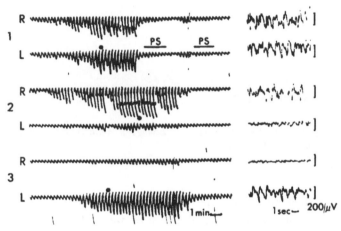

Fig. 2. Three sleep episodes in the brain hemispheres of a northern fur seal. Left: integrated EEG (epoch of integration 20 sec, time constant 0.2 sec). Right: examples of EEG from the epochs marked by black dots. R, right hemisphere; L, left hemisphere; PS, paradoxical sleep episodes. 1, bilateral slow wave EEG; 2, right-side EEG synchronization; 3, left-side EEG synchronization.

which slow wave EEG patterns are recorded in one hemisphere (alternatively in the right or the left one) while EEG-desynchronization is recorded simultaneously in the other hemisphere (Figs. 1 and 2). In addition, there are frequent episodes particularly in fur seals, when both hemispheres clearly show two different forms of synchronization, such as delta waves of maximum amplitude in one hemisphere and slow waves of low amplitude in the other. Such a state of distinct interhemispheric asymmetry of slow wave EEG cannot be regarded as unihemispheric slow wave sleep in the strict sense. Nevertheless, we are using the same term for both, hoping that further experimentation will provide new data and enable us to better comprehend the mechanism and functional role of this phenomenon; then our terminology may become more accurate.

The results summarized in this review have been obtained by the same group of researchers during the last 10 years. Sleep was studied electrophysiologically in 23 adult bottlenose dolphins, 9 adult porpoises, 17 northern fur seals of different age, 4 adult Caspian seals, and 6 harp seal whelps. Continuous polygraphic recordings for up to 7 days were made on animals which were freely swimming in experimental tanks or in open-sea pens. Pinnipeds' sleep was studied not only in the water but also on the land.

I. PHENOMENOLOGY OF UNIHEMISPHERIC SLOW WAVE SLEEP

Unilateral slow wave EEG synchronization has been revealed in all bottlenose dolphins, porpoises and fur seals irrespective of sex and age, provided that they had implanted electrodes in both hemispheres. However, it was not found in any of the Caspian and harp seals. These observations suggest that the presence or absence of unihemispheric sleep is a species characteristic of marine mammals. Since the number of Cetacea and Pinnipedia species studied is rather small, it is so far difficult to say which of their ecological or morphophysiological features correlate with the presence of unihemispheric sleep.

While in dolphins intermediate synchronization (sleep spindles and low theta and delta waves) can be both bilateral and unilateral, the delta stage (delta waves of maximum amplitude with delta index over 50%) can be only unilateral. This suggests reciprocal relations between the two hemispheres of the dolphin during delta sleep generation. In fur seals, both the intermediate synchronization and the delta stage can be unilateral, like in dolphins, and bilateral, like in terrestrial mammals. One may attribute this difference in delta sleep organization between dolphins and fur seals to the fact that the former sleep only in the water, while the latter sleep in the water and on the land. However, it is noteworthy that unilateral slow waves can be recorded in fur seals not only when they sleep in the water but also on land.

In dolphins, intermediate bilateral synchronization, that corresponds to ordinary superficial slow wave sleep, ranges from 2 to 10% of total recording time in different individuals. All variants of unihemispheric slow wave sleep amount to about 30 to 40% of recording time. No paradoxical sleep was found in dolphins, as has been mentioned in our recent review (5). Thus, the unihemispheric form is the main type of dolphins' sleep. In fur seals which are 2 to 3 years old, total slow wave sleep occupies nearly 30% of the day. About half of that time the slow waves in the two hemispheres are asynchronous. In other words, unihemispheric slow wave sleep in fur seals is present for a considerable time. Paradoxical sleep in fur seals accounted for about 6% of recording time (6).

Usually, unihemispheric sleep episodes in dolphins last some dozens of minutes. The maximum length of a single continuous, unilateral slow wave episode recorded in bottlenose dolphins was 2.5 hr. Asynchronous slow wave sleep episodes in fur seals are much shorter than in dolphins, lasting usually for several minutes. In most cases, unihemispheric slow waves in fur seals are observed at the beginning of slow wave sleep episodes which later become bilateral.

After having discovered the unilateral slow waves, we investigated the phenomenon more closely to see whether it consisted of i) interhemispheric differences, ii) local EEG synchronization, or iii) desynchronization which can only occur within one cortical field in one of the hemispheres. We carried out test experiments on bottlenose dolphins and fur seals whose electrical activity in frontal, parietal, occipital, and temporal areas of both hemispheres was simultaneously recorded. It turned out that in this case interhemispheric differences were present rather than local EEG synchronization or desynchronization in a certain cortical field. Although the onset and end of a slow wave episode may not always coincide in different areas of one hemisphere, each hemisphere acts as a unit during unihemispheric sleep.

As seen from records of different thalamic nuclei in bottlenose dolphins, unihemispheric slow wave sleep is not only a cortical, but also a subcortical phenomenon. The slow waves registered in the thalamus could also be unihemispheric, and evolved always synchronously with slow wave activity of the ipsilateral neocortex.

Cortical temperature of the dolphin's brain also reflects unihemispheric slow wave sleep organization (3). The temperature decreases with EEG synchronization and increases with desynchronization. However, these temperature changes occur in both hemispheres independently and in strict accordance with the changes of the electrical activity in the same hemisphere.

In bottlenose dolphins, the total amount of slow wave sleep in one hemisphere was found to be sometimes considerably larger than in the other (up to two-fold). This inequality was observed from one daily session to another. Hence, in bottlenose dolphins one hemisphere sleeps more than the other. In some individuals this happened to be the right hemisphere, in others the left one. No interhemispheric asymmetry of this kind was found in fur seals in whom the total amount of sleep in the two hemispheres showed no large difference. In various daily sessions one of the hemispheres was observed to sleep sometimes more and sometimes less than the other.

Observing the dolphins we did not notice any asymmetry in movement or posture coinciding with unilateral slow wave EEG. A peculiar feature of the dolphin's behavior is the frequent unilateral closure of an eye. However, as reported earlier (14), there is no strict correlation between alternating eye opening or eye closing, and alternating sleep of the hemispheres. If the dolphin has only one eye open, the electrical activity of the contralateral hemisphere may be desynchronized or synchronized. This is particularly surprising in view of the fact that the optic nerve fibers in the optic chiasm of the bottlenose dolphin decussate completely (2). During the dolphin's sleep its open eye performs a sentinel function: when the visual field of the open eye is noiselessly exposed to an unusual pattern the dolphin is behaviorally awakened, regardless of whether its contralateral or ipsilateral hemisphere has been asleep. Thus, unlike in terrestrial mammals in which the auditory system alone maintains contact with the environment during sleep, the visual system in dolphins performs a similar function.

Northern fur seals are immobile during sleep on land. When in the water, they sleep on the water surface with their nostrils above water. They maintain their special sleep posture by moving only one of their front flippers. The asymmetric EEG in fur seals sleeping in the water may be attributed to the asymmetric motor activity of their flippers. However, a statistical analysis of this correlation has not yet been made.

Another behavioral peculiarity of dolphins is their sleep during for movement. As noted by McCormick (4) for *Phocoenoides dalli*, "there has never been any observation of activity resembling sleep behavior." No behavioral immobility was found in river dolphins either (*11*). According to our observations, porpoises *Phocoena phocoena* swim all day long without interruption, so that all forms of their sleep occur on the background of constant movement. Some authors assumed that dolphins sleep only during respiratory pauses, and that they must wake up when they emerge for respiration. As found in our electrophysiological experiments, dolphins may wake up during respiration, but not in all cases. When they are in the state of intermediate bilateral sleep or in any stage of unihemispheric sleep, the dolphin's movements related to its respiration may not be accompanied by bilateral desynchronization.

In bottlenose dolphins, sleep during constant swimming is also possible. However, unlike porpoises, bottlenose dolphins may be easily observed hovering near the water surface without swimming. Most observers attributed this behavioral state to sleep. However, even while hovering, the dolphin's fins continue moving and provide thereby a stable posture. Our polygraphic records demonstrated that when bottlenose dolphins hover at the water surface wakefulness is as possible as sleep.

Thus, behavioral immobility is not a compulsory condition for dolphins' sleep. This assumption contradicts the widespread idea that the biological function of sleep is to provide immobility for the sake of energy saving.

II. MECHANISM AND FUNCTIONAL ROLE OF UNIHEMISPHERIC SLOW WAVE SLEEP

So far, the neurophysiological mechanism of unihemispheric sleep in dolphins and fur seals is still unknown. Our working hypothesis is that the right-side and left-side parts of some desynchronizing and synchronizing systems in the lower parts of the brainstem are functionally disintegrated and interact reciprocally.

It is of great interest to find out how the unihemispheric sleep mechanism is compatible with the hypothesis of a humoral control of sleep. It is important to note that the two brain hemispheres of dolphins and fur seals have a common blood supply. In view of this, the independent sleep in the two hemispheres contradicts the idea that humoral factors play an important role in sleep regulation. The same conclusion has been reported in the literature with respect to the independent sleep of conjoined twins' heads with a common circulatory system (15). In the case of unihemispheric sleep in dolphins, the situation is still more convincing, because, independently of each other, not two entire brains but two hemispheres of one and the same brain exhibit sleep. This opinion has been confirmed in our experiments with pentobarbital injections which provoked in dolphins only bilateral EEG slow waves.

On the other hand, diazepam injections provoked unihemispheric sleep in dolphins. Hence, generally speaking, there may exist humoral factors which not only block the mechanism of unihemispheric EEG synchronization, but even favor this process.

Our experiments with drug injections to dolphins have demonstrated that the unihemispheric sleep is connected with respiration. The experiments have shown dolphins' autonomous breathing to be incompatible with bilateral delta-waves. When a pentobarbital injection provokes bilateral delta waves of maximal amplitude, dolphins' autonomous respiration stops. Diazepam injections occasionally provoked bilateral delta-waves in dolphins, but only during respiratory pauses. Before a respiration the EEG of one or both hemispheres always become desynchronized. In this latter case, the typical pattern of unihemispheric slow wave sleep was observed (Fig. 3).

All marine mammals face the problem of combining their sleep in the aquatic environment with the necessity to move for obtaining atmospheric air for breathing. Marine animals seem to have solved this problem in different ways. For example, Caspian seals sleep during the

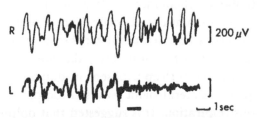

Fig. 3. Unilateral electrocortical arousal coinciding with the respiration in a porpoise after diazepam administration (1 mg/kg i.v.). R, right hemisphere EEG; and L, left hemisphere EEG. Bipolar recordings from roughly symmetrical areas of the parietal cortex. Solid line marks an expiration-inspiration act of the dolphin.

respiratory pauses and arouse for the respiration act which necessitates movement for floating up or raising the head above water level. When they do not need to move for breathing, which happens during their sleep on land or on the quiet surface of the sea, Caspian seals can sleep continuously and remain unaroused for breathing. Under such conditions, their sleep resembles sleep in terrestrial mammals. It seems likely that dolphins have adapted differently to sleep in the water. By evolving the mechanism of unihemispheric slow wave sleep, one hemisphere is able to sleep for a comparatively long time without short-term arousals, while the other remains sufficiently active to preserve high muscle tone and high reflectory activity of the brain, so that respiration can proceed normally and safely. The dolphin's sleep pattern in the water may be also attributed to the necessity to remain in continuous movement not only for the respiration acts as in the Caspian seal. It is reasonable to assume that the absence of paradoxical sleep in dolphins and the evolution of unihemispheric delta sleep have a common origin.

SUMMARY

Unihemispheric slow wave sleep has been demonstrated in the bottlenose dolphin, porpoise, and northern fur seal; it was not found in Caspian and harp seals. Intermediate EEG synchronization in dolphins can be both bilateral and unilateral, while the delta stage can be only unilateral. In fur seals, both the intermediate synchronization and the delta stage can be bilateral and unilateral. The interhemispheric asynchrony of slow waves does not result from local EEG synchronization or desynchronization. Unihemispheric slow wave sleep in dolphins is not only a cortical, but also a thalamic phenomenon. It is manifested not only by the EEG but also by the fluctuations of cortical brain temperature. We failed to find any correlates between unihemispheric sleep and behavior in dolphins. In any of its behavioral state, the dolphin exhibits active fin movement. Paradoxical sleep is absent in dolphins, but is well documented in pinnipeds. Bilateral delta waves in dolphins are incompatible with autonomous respiration. It is suggested that dolphins need unihemispheric sleep to maintain the motor activity which is necessary for normal autonomous respiration.

REFERENCES

1 Huntley, A.C., Costa, D.P., Walker, J.M., Walker, L.E., and Berger, R.J. (1981). APSS 21th Annual Meeting, Hyannis.
2 Jacobs, M.S., Morgane, P.J., and McFarland, W.L. (1975). *Brain Res.* **88**, 346–352.
3 Kovalzon, V.M. and Mukhametov, L.M. (1982). *J. Evol. Biochem. Physiol.* **18**, 307–309 (in Russian).
4 McCormick, J.G. (1969). *Proc. Natl. Acad. Sci. U.S.* **62**, 697–703.
5 Mukhametov, L.M. (1984). *Exp. Brain Res.* (Suppl. 8), 227–238.
6 Mukhametov, L.M., Lyamin, O.I., and Polyakova, I.G. (1984). *J. High. Nerv. Act.* **34**, 465–471 (in Russian).
7 Mukhametov, L.M. and Polyakova, I.G. (1981). *J. High. Nerv. Act.* **31**, 333–339 (in Russian).
8 Mukhametov, L.M., Supin, A.Y., and Polyakova, I.G. (1977). *Brain Res.* **134**, 581–584.
9 Mukhametov, L.M., Supin, A.Y., and Polyakova, I.G. (1984). *J. High. Nerv. Act.* **34**, 259–264 (in Russian).
10 Mukhametov, L.M., Supin, A.Y., and Strokova, I.G. (1976). *Proc. USSR Acad. Sci.* **229**, 767–770 (in Russian).
11 Pilleri, G. (1979). *Endeavour* **3**, 48–56.
12 Ridgway, S.H., Harrison, R.J., and Joyce, P.L. (1975). *Science* **187**, 553–555.
13 Sokolov, V.E. and Mukhametov, L.M. (1982). *J. Evol. Biochem. Physiol.* **18**, 191–193 (in Russian).
14 Supin, A.Y., Mukhametov, L.M., Ladygina, T.F., Popov, V.V., Mass, A.M., and Polyakova, I.G. (1978). *Electrophysiological Study of Dolphins Brain*. Moscow: Nauka (in Russian).
15 Webb, W.B. (1978). *Sleep* **1**, 205–211.

7

SLEEP AS AN ADAPTIVE PROCESS

SHIZUO TORII

Department of Physiology, Toho University School of Medicine, Tokyo 143, Japan

Since slow wave sleep (SWS) is the deepest stage of non-REM (NREM) sleep, the amount of SWS is thought to be an indicator of sleep quality. Recent studies, however, showed that benzodiazepine abolished SWS and increased stage 2 sleep, accompanied by the subjective estimation of better sleep quality (4). Furthermore, experimental deprivation of SWS (stage 4 sleep) has not produced a substantial behaviour deficit (6). These observations raise the question whether SWS is essential or not. In this context, nocturnal sleep patterns of nurses recorded in our laboratory are worthwhile to consider (12). The experienced nurses reported better sleep quality in spite of a reduced amount of SWS, while the inexperienced nurses reported poor sleep quality even though they had more SWS. Thus, it is likely that the nurses who are forced to have an irregular short sleep under the three shift system show a diminution of need for SWS which therefore may be considered an adaptive response for their enforced short sleep. Changes in sleep stages of the nurses suggest that there is some plasticity with respect to SWS.

I. SLEEP OF SHIFT WORKERS

Results of experiments using polysomnography have shown that the day sleep of shift workers somehow deteriorates (1, 2, 5, 8). The findings of these studies suggested that shift workers show 1) a reduction of sleep time; 2) an increase in awakenings after sleep onset and in stage 1 sleep; and 3) a decrease in stage 4 sleep and REM sleep. The factors which caused sleep disturbance of shift workers, especially night workers, are considered to be due to the desynchronization of the circadian rhythm of various physiological functions (2, 10). Prolonged night work causes a person to develop chronic sleep disturbances because human beings cannot adapt completely to night work. Therefore, it is advisable to adopt a rapid rotation of shifts or a single night shift as a preferable schedule (10). From this point of view, a number of hospital nurses have been adopting the work-sleep schedule so that they can maintain the habit of sleeping at night, without working consecutively in the evening or at night, and having day work as the basis. When the day shift is followed by the night shift, most of the nurses are taking a night nap before the night work, and after the night work they are taking a day nap as well as the night sleep.

However, even when adopting such a work-sleep schedule, it is very common for the nurses to have irregular short sleep when day work is followed by night work or when evening work is followed by day work (8). Until now, the following questions remain unanswered. How does irregular short sleep influence the quality of the night sleep of nurses? How does this kind of sleep-work schedule affect the night sleep if it is continued for many years?

To study the characteristics of night sleep of nurses having irregular short sleep, they were compared with a control group of female students. Four nurses (21–26 years old) and 4 female students (20–22 years old) took part in this study.

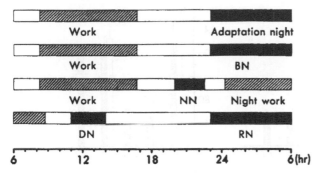

Fig. 1. Schematic diagram of experimental design. Consecutive 24 hr periods are plotted below each other starting with the adaptation day. ▨ work; ▮ sleep. BN, baseline night; NN, night nap; DN, daytime nap; RN, recovery night.

II. EXPERIMENTAL DESIGN TO CLARIFY SLEEP CHARACTERISTICS OF NURSES

According to the work-sleep schedule of the nurses, the following four sleep conditions were employed in the experiment (Fig. 1).

1) Baseline night sleep (from 23:00 to 6:00)
2) Night nap before night work (from 20:00 to 22:30)
3) Daytime nap after night work (from 11:00 to 14:00)
4) Recovery night sleep (23:00 to 6:00)

The nurses were engaged in their usual work and the students who took part in the experiment spent their time as usual (attending classes in the daytime). To make their conditions similar to those of the nurses, the students were asked to stand for 50 min and to sit for 10 min in each of the 8 hr between the night nap and the daytime nap.

Sleep was recorded by polysomnography in each sleep condition in a darkened, sound-attenuated and air-conditioned room. The records were classified according to Rechtschaffen and Kales (9) for one minute periods.

III. DIFFERENCES BETWEEN NURSES AND STUDENTS IN THE BASELINE NIGHT

Among the sleep variable indices, the sleep efficiency index and the number of awakenings did not differ between the groups, while the

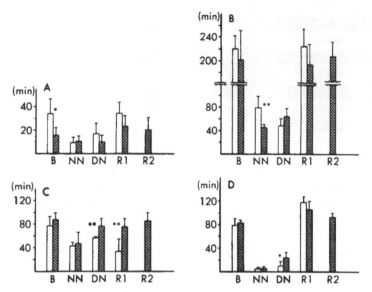

Fig. 2. Amount of each sleep stage (mean with S.E.M.) for all 4 sleep conditions (see Fig. 1). A: stage 1, B: stage 2, C: stage 3+4. D: REM sleep. ☐ nurses; ▨ students. *$p<0.05$, **$p<0.02$.

number of stage 1 and number of stage shifts increased in the nurses as compared to the students. As to the amount of sleep stages, the nurses had more stage 1, more stage 2 sleep and less SWS. There was no significant difference between the students and the nurses in the amount of REM sleep, although REM latency was significantly longer in nurses than in students ($p<0.05$). This was due to the absence of the initial REM sleep episode in three nurses.

These sleep changes led us to the notion that the nurses have sleep disturbances. However, it should be pointed out that the distributions of stage 1 and stage 3+4 sleep throughout the night are quite different from those typical for a sleep disturbance.

Nurses had more stage 1 sleep towards the end of the sleep period and more stage 3+4 sleep in the beginning of the sleep period. As shown in Fig. 3, stage 4 sleep of nurses remains confined to the first third of the sleep period, while stage 4 sleep of students tends to extend to the last third of the sleep period. The temporal distribution of stage 4 and stage 1 sleep suggests that nurses sleep quite efficiently, because stage 4 sleep is not interrupted by stage 1 sleep. Furthermore, there was less

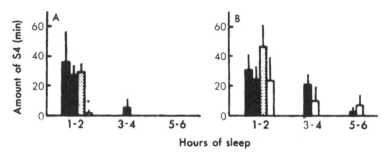

Fig. 3. Amount of stage 4 (S4) sleep for consecutive 2 hr periods after sleep onset plotted for each sleep condition. A: nurses. B: students. *p< 0.05. ▮ BN; ▨ NN; ▨ DN; ▢ RN.

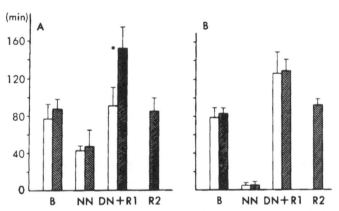

Fig. 4. Comparison of amount of stage 3+4 and REM sleep during the 24 hr period. A: stage 3+4. B: REM sleep. ▢ nurses; ▨ students. *p<0.05.

sleep interruption and nocturnal waking in the nurses which was different from observations of daytime sleep of night workers (*1, 2, 5, 7, 8*) and sleep disorders in elderly persons (*14*). Thus, it is not likely that the night sleep of nurses is disturbed.

IV. STAGE 4 SLEEP REBOUND IN NURSES AND STUDENTS

There was no significant difference in the amount of stage 4 sleep in the baseline night between nurses and students (Fig. 4). In the recovery night, however, students had more stage 4 sleep than nurses.

On the 2nd experimental day the subjects took a night nap for 3 hr

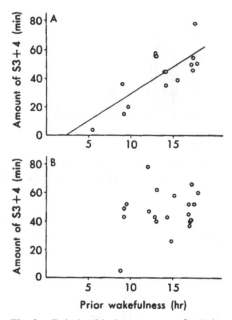

Fig. 5. Relationship between stage 3+4 sleep (S3+4) in the first 2 hr of each sleep condition and prior wakefulness. A: nurses; $r=0.77$. B: students; $r=0.21$.

instead of a night sleep. This situation can be regarded as a partial sleep deprivation (*3, 13*). On the 3rd day following partial sleep deprivation, the subjects took both a daytime nap and a night sleep. The total amount of stage 3+4 sleep on the 3rd day was expected to be much higher than on the baseline night. It is striking that stage 3+4 sleep in nurses did not increase significantly on the 3rd day, whereas stage 3+4 sleep in students did increase significantly.

If the compensatory increase of stage 3+4 sleep after partial sleep deprivation indicates a need for SWS, it follows that the nurses of this study showed a reduced need for SWS. This assumption is supported by the stage 4 response to prior wakefulness. It is well documented that the stage 4 response is a function of the length of prior wakefulness. This relationship was examined for the first 3 hr of each sleep condition. A significant correlation was noted in the nurses, whereas no clear relation was found in the students (Fig. 5). In other words, the stage 4 response to prior wakefulness was higher in students than in nurses.

Since stage 2, 3, and 4 sleep appears to be part of a single state

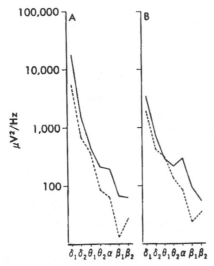

Fig. 6. Frequency distribution of the EEG power density in the daytime naps for SWS and stage 2 sleep. A: SWS; $p<0.001$ at each point ($n=55$). B: stage 2; $p<0.001$ at each point except for θ_1 ($n=55$). —— students; - - - - nurses. δ_1, 0.25–2.00; δ_2, 2.25–4.00; θ_1, 4.25–6.00; θ_2, 6.25–8.00; α, 8.25–13.00; β_1, 13.25–19.00; β_2, 19.25–64.00 Hz.

which is differentiated arbitrarily into three stages on the basis of the amount of delta activity, it is possible that the decrease in the number of delta waves with stage 4 diminution is compensated by increased delta activity in stage 2 sleep. To examine this point, the EEG records from the daytime nap in which the diminition of stage 4 sleep in nurses was the most marked, were used to compute the frequency distribution of the power density in the 0.25–64.0 Hz range for the visually scored sleep stages.

The frequency distribution of the EEG power density in the daytime nap was plotted for stage 2 sleep and SWS (Fig. 6). The curves connect the mean values (4 subjects each) computed for 0.25 Hz bins. It can be seen that the predicted compensatory changes in delta activity in nurses occurred neither in stage 2 sleep nor in SWS.

We can summarize the sleep patterns of nurses compared with those of female students as follows: 1) most of SWS appeared in the early part of the sleep period; 2) stage 1 sleep increased in the later half of the sleep period; and 3) SWS rebound was less marked.

V. DIFFERENCES BETWEEN EXPERIENCED AND INEXPERIENCED NURSES IN THE BASELINE NIGHT

Since female students do not appear to be a suitable control group in this study, experienced nurses (their length of experience is more than 3 years) were compared with inexperienced nurses (their length of experience is less than 3 months) under the same experimental design. Inexperienced nurses had longer sleep onset latency, more nocturnal awakenings, a reduced sleep efficiency index, and more stage shifts (especially shift to stage 1 from stage 2, 3, and 4 sleep was increased significantly) which are all considered as indicators of sleep disturbance. Also, their subjective estimation of sleep in the recovery night was examined. The experienced nurses reported a good sleep quality, while the inexperienced nurses reported poor sleep quality in spite of more sleep. These data suggest that sleep of the experienced nurses may be interpreted as more "efficient." Our data imply that there is some plasticity with respect to SWS and that by its adaptation it can conform to a changed life style.

SUMMARY

To examine whether sleep is an adaptive process or not, the night sleep of female students, inexperienced nurses and experienced nurses was compared. Sleep patterns of nurses showed a different pattern from that of irregular sleepers or short sleepers. Furthermore, the nurses' sleep patterns were quite different from a sleep disorder pattern. The sleep structure of experienced nurses could be considered to reflect the adaptation to an irregular short sleep.

REFERENCES

1 Bryden, G. and Holdstock, T.L. (1973). *Psychophysiology* 10, 32–42.
2 Dahlgren, K. (1981). *Psychophysiology* 18, 381–391.
3 Dement, W. and Greenberg, S. (1966). *Electroenceph. Clin. Neurophysiol.* 20, 523–526.
4 Feinberg, I., Fein, G., Walker, J.M., Price, L.J., Floyd, T.C., and Narch, J.D. (1977). *Science* 198, 847–848.
5 Foret, J. and Benoit, O. (1978). *Waking Sleep.* 2, 107–117.

6 Horne, J.A. (1978). *Biol. Psychol.* **7**, 55–102.

7 Kripke, D.F., Cook, B., and Lewis, O.F. (1971). *Psychophysiology* **7**, 377–384.

8 Matsumoto, K. (1978). *Waking Sleep.* **2**, 169–173.

9 Rechtschaffen, A. and Kales, A. (1968). A Manual of Standard Terminology, Techniques and Scoring System for Sleep Stages of Human Subjects. Washington D.C.: Public Health Service U.S. Government Printing Office.

10 Rutenfranz, J., Knauth, P., and Colquhoun, W.P. (1976). *Ergonomics* **19**, 331–340.

11 Taub, J.M. and Berger, R.J. (1973). *Electroenceph. Clin. Neurophysiol.* **35**, 613–619.

12 Torii, S., Okudaira, N., Fukuda, H., Kanemoto, H., Yamashiro, Y., Akiya, M., Nomoto, K., Katayama, N., Hasegawa, M., Sato, M., Hatano, M., and Nemoto, H. (1982). *J. Human Ergol.* **11** (Suppl.), 233–244.

13 Webb, W.B. and Agnew, H.W. Jr. (1965). *Science* **150**, 1745–1746.

14 Williams, R.L., Karacan, L., and Hursch, C.J. (1974). *Electroencephalogram (EEG) of Human Sleep: Clinical Application.* New York: John Wiley and Sons.

CIRCADIAN RHYTHM

8

CIRCADIAN REST-ACTIVITY CYCLE IN RATS: ITS MODULATION BY HUMORAL FACTORS AFFECTING THE CENTRAL NEUROTRANSMITTER SYSTEMS

KEN-ICHI HONMA

Department of Physiology, Hokkaido University School of Medicine, Sapporo 060, Japan

In recent sleep research, there is an increasing tendency to view sleep and associated phenomena in relation to the circadian oscillation. The circadian rhythms of nocturnal rodents are regulated by a self-sustained pacemaker(s) which is located in the brain. Functions of the pacemaker (*e.g.*, entrainment to the light cycle or coupling to many overt rhythms) are thought to be mediated by humoral factors such as neurotransmitters and brain peptides, but the central mechanism is not well understood. There are a number of neurochemical substances which affect sleep and wakefulness (*22, 27*), but only very few which alter the sleep-wakefulness cycle or the circadian oscillation (*11*). In the present paper, a neuropharmacological approach was used to investigate the relation between the pacemaker functions and the central neurotransmitter systems. Since the function of the circadian pacemaker can be assessed only by studying the overt rhythm of some variables, changes in such rhythm studied do not necessarily indicate changes in the pacemaker. The intrinsic period and phase of a circadian rhythm are the most reliable indicators of the pacemaker. The shape of the rhythm (amplitude or activity time) is sometimes useful to speculate what happens in the circadian pacemaking system. To study the pacemaker functions over a long time period, the spontaneous locomotor activity was measured, the daily time course of which is known to be almost identical to that of the sleep-wakefulness cycle monitored by electroencephalogram (EEG) (*3*).

I. THE PACEMAKER STRUCTURE OF CIRCADIAN RHYTHM IN RATS

Before focussing on the brain mechanism of the circadian rhythm, the structure of the pacemaking system in rats and the interrelationship of free-running period (τ), activity time (α) and sensitivity of the pacemaker to light are briefly discussed. The sensitivity to light is expressed as a phase response curve (PRC). The phase responsiveness is the most important property of the circadian pacemaker and is a basic feature of the entrainment to the light cycle (*23*).

Four different PRCs were established by applying single light pulses (300 lux, 30 min duration) every second week to free-running rats under constant darkness (DD) (*16*). Phase shifts of the circadian locomotor rhythm produced by the light pulses were plotted in reference to the circadian time (CT) when the light pulse was given. Both the onset and offset of activity band in the circadian rhythm were used as a phase-reference (onset PRC and offset PRC). The onset and offset PRCs were different in shape, in particular, when the phase shifts on the day following the light pulse were measured (immediate PRC). The advance (A) area in the PRC, where the light pulse produces an advance phase shift, was lacking in the onset PRC, whereas the delay (D) area, where a delay phase shift occurs in response to the pulse, was observed in both PRCs. The difference in the PRC shape was less clear when phase shifts were measured after a new steady state was achieved (steady state PRC). A lack of the A area was also reported in the PRC constructed by using the activity of rat pineal enzyme, serotonin N-acetyl-transferase, as the indicator (*18*). These findings support a two-oscillator pacemaking model proposed by Pittendrigh and Daan (*25*). The onset and offset components of the locomotor activity are separately driven by two mutually coupled oscillators, whose sensitivities to light are different. Furthermore, the PRC shape depends on τ in DD and on α just prior to the light pulse (16). The range covered by the D area in the PRC was longer in the long τ or α rhythms than in the short τ or α rhythms. In other words, the longer τ or α, the larger the D/A ratio.

From this interdependence between the PRC shape on one hand, and, τ and α on the other hand, it is hypothesized that τ and α are related to the coupling strength which determines the phase relation between the two oscillators comprising the circadian pacemaker of rats.

II. ENTRAINMENT TO LIGHT

One of the most important roles of the circadian system is to keep a stable phase relation between an overt rhythm and an external cycle, especially the light-dark (LD) cycle, and to maintain a temporal order among many physiological functions within an organism. As mentioned, the entrainment is based on a periodically changing sensitivity of the pacemaker to an entraining signal, which is expressed as the PRC. A single light pulse is an effective stimulus but we do not know the mechanism which mediates the light signal to the circadian pacemaker in the body.

Carbachol, a cholinergic agonist, was reported to mimic the effects of light in suppressing an enzyme activity in the rat pineal, and in

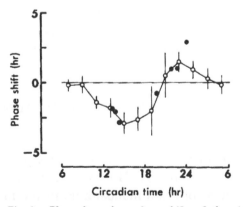

Fig. 1. Phase-dependent phase shifts of the circadian locomotor rhythm induced by intraventricular injections of carbachol (10 μg) to blinded rats. Closed circles indicate the phase shifts by carbachol. Open circles with solid lines indicate the onset PRC (means of 2-hr bins and SD) obtained by single light pulses.

phase-shifting the free-running locomotor rhythm of mice (*32, 33*). Figure 1 demonstrates the phase-dependent phase shifts of the circadian locomotor rhythm induced by intraventricular injections of carbachol to blinded rats. The phase shift was measured after a steady state was achieved. The onset PRC by single light pulses was superimposed in the figure for the purpose of comparison. There is a marked tendency for carbachol injected during the early subjective night to produce a phase delay shift, and during the late subjective night to produce a phase advance shift. The relation between the direction of phase shift and the phase of circadian cycle at which it occurs was quite similar to the PRC obtained by a single light pulse. In addition, carbachol was reported to simulate the effects of light in the photoperiodic control of reproduction in hamsters (*8*).

The entraining light signal is mediated by the retinohypothalamic tract terminating in the suprachiasmatic nucleus (SCN). Although the neurotransmitter involved in the retinohypothalamic projection is unknown, choline acetyltransferase, a synthetic enzyme of acetylcholine (*6*), and the nicotinic cholinergic receptor (*28*) have been demonstrated in the SCN. These results suggest that a central cholinergic mechanism is involved in the entrainment to the LD cycle. Recently, avian pancreatic polypeptide (APP) was shown to have a phase dependent phase-shifting effect in rats (*2*). Contrary to the effects of light pulse or carbachol injection, APP administration in the early subjective night induced an advance phase shift, and in the late subjective night a phase delay shift. The effect of APP was similar to dark pulses applied in constant illumination (*5, 9*). More than one humoral factor seems to be involved in the entrainment to light.

III. THE PERIOD AND SHAPE OF THE FREE-RUNNING CIRCADIAN RHYTHM

It is well established that central monoaminergic mechanisms play an important role in the regulation of sleep (*12*). However, there is little evidence suggesting an involvement of brain monoamines in the function of the circadian pacemaker. Serotonergic nerve endings are found in the SCN, whose cell bodies are located in the dorsal raphe nucleus (*1*). The uptake of serotonin by brain tissue containing the SCN showed a 24-hr periodicity (21). Reduction of the brain serotonin level by parachlorophenylalanine (PCPA), an inhibitor of a serotonin synthetic enzyme, or by 5,6- or 5,7-dihydroxytryptamine (DHT), agents which destroy the serotonergic nerve terminals permanently, only produced transient effects on circadian rhythms of locomotor activity or plasma hormone levels (14, 19). Similarly, depletion of the central catecholamines by α-methyl-p-tyrosine, or by 6-hydroxydopamine (6-OHDA) had a limited effect on the circadian oscillation (13, 29). Figure 2 illustrates the effects of intraventricular injection of 6-OHDA and 5,6-DHT on the circadian locomotor rhythm of rats in LD. 6-OHDA reduced the amplitude of locomotor activity during the dark period

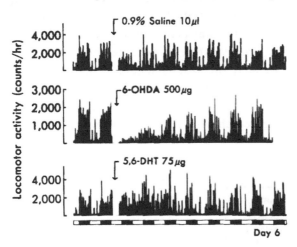

Fig. 2. Effects of intraventricular 6-OHDA (500 μg) and 5,6-DHT (75 μg) injections on the circadian locomotor rhythm of rats in LD (LD 12:12). Black parts of the horizontal bar at the bottom indicate the 12-hr dark period.

TABLE I
Free-running Parameters (Level, Amplitude, and Period) of the Circadian Locomotor Rhythm of 6-OHDA, and 5,6-DHT Treated Rats in 200 lux LL.

	Level	Amplitude (counts/15 min)	Period (hr)
Saline (6)	487±38	319±31	25.1±0.1
6-OHDA (6)	456±55	284±54	25.0±0.2
5,6-DHT (6)	496±73	287±37	25.4±0.1

The values are expressed by the mean and S.E. The numbers in parentheses indicate numbers of rats examined. The parameters were calculated by a modified method of cosinor analysis.

and 5,6-DHT enhanced the amplitude during the light period. The effects, however, lasted only for a few days, even though the brain monoamines level remained low. Then these animals were transferred to constant illumination (LL), and the free-running period, amplitude and level of the circadian rhythm were measured (Table I). There was no significant difference in these parameters between the amine depleted and control groups.

These results indicate that neither the catecholaminergic nor the serotonergic mechanism is involved in the pacemaker functions. However, a prolonged depletion of the central monoamines may cause an increase in the turnover rate of amine metabolisms, and/or a change in postsynaptic mechanisms (*e.g.*, sensitivity to the neurotransmitter). Furthermore, we do not known what happens to the circadian pacemaker during the first days after drug treatment, because the overt circadian rhythm was disrupted. In fact, the results suggest a shortening of the free-running period during the initial days of PCPA treatment (*13*). Figure 3 illustrates the circadian locomotor rhythm of a rat treated with PCPA in LL. PCPA decreased the brain serotonin level for more than

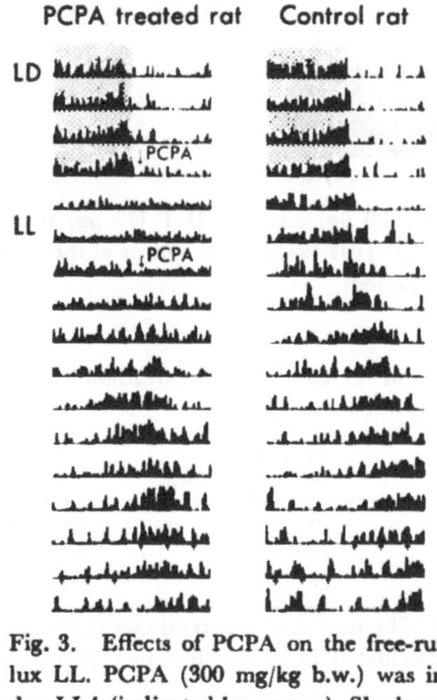

Fig. 3. Effects of PCPA on the free-running locomotor rhythm of rats exposed to 200 lux LL. PCPA (300 mg/kg b.w.) was injected intraperitoneally twice on day LL1 and day LL4 (indicated by arrows). Shadowed area indicates the dark period.

TABLE II
The Phase on LL11 (day 11 of LL) and Free-running Parameters (Period, Amplitude, and Level) of the Circadian Locomotor Rhythm of PCPA Treated Rats in 200 lux LL.

	Acrophase on LL11 (hr: min)	Free-running parameters after LL11		
		Level (counts/ 15 min)	Amplitude (counts/ 15 min)	Period (hr)
Saline (6)	13:44±45	339±17	197±24	25.3±0.1
PCPA (6)	08:45±38*	384±19	179±46	25.1±0.1

Values are expressed by the mean and standard error. Numbers in parentheses indicate numbers of rats examined. The parameters were calculated by a modified method of cosinor analysis. Asterisk indicates a statistically significant difference between the saline and PCPA treated groups.

a week and abolished the circadian rhythm. After the rhythm reappeared, the phase, free-running period, amplitude and level of the circadian rhythm were measured and compared with those of the saline injected control group. The results are given in Table II. The acrophase on day LL11, a few days after the reappearance of the circadian rhythm, was significantly advanced in the PCPA treated group as compared to the control group. The other parameters of the circadian rhythm did not differ significantly between the two groups. A similar phase advance of the circadian rhythm was observed in the plasma corticosterone level of PCPA treated rats. The difference in phase position after PCPA treatment suggests a change in τ during the period when the overt circadian rhythm was abolished by PCPA. More studies of well-designed experiments are needed to elucidate the roles of the serotonergic system in the function of the circadian pacemaker.

IV. INTERNAL COUPLING OF THE CIRCADIAN LOCOMOTOR RHYTHM

A unique phenomenon termed splitting provided the first circumstantial evidence for a two-oscillator pacemaking model in rodents (24). The circadian locomotor rhythm in rodents split into two activity components which kept a phase difference of almost 180 degrees for a long period. Splitting occurs spontaneously in LL. Figure 4 illustrates an example of splitting observed in the circadian locomotor rhythm of a rat exposed to LL. In rats splitting does not persist for a long time, and the two separated activity components soon fuse again (4, 7). As de-

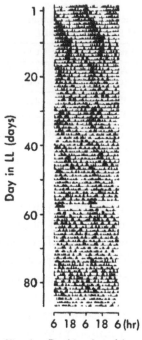

Fig. 4. Double-plotted locomotor rhythm of a rat exposed to 200 lux LL. 2 weeks after the exposure to LL, the circadian rhythm split into two activity components but the split components were fused again.

monstrated in the figure, splitting and refusion recurred several times. Splitting can be explained as a result of an altered internal coupling between two oscillators which comprise the circadian pacemaker.

The circadian organization of the locomotor activity in hamsters was reported to be destroyed by continuous administration of either imipramine, a tricyclic anti-depressant drug, or clorgyline, a specific inhibitor of monoamine oxidase (31). D-Amphetamine, a potent releaser of catecholamines, had a similar effect on the circadian rhythm of rats (17). To confirm and extend the latter finding, methamphetamine was dissolved in drinking water at the concentration of 0.01% and administered to rats for a period of more than 6 months (15). In the first two months of drug treatment, the circadian locomotor rhythm remained essentially unaltered, except for a slight augmentation of the activity during the dark period. Mild anorexia was observed in all rats, but not to the extent of body weight loss. After 2 months of treatment, the phase

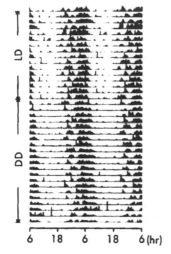

6 18 6 18 6 (hr)

Fig. 5. Double-plotted circadian locomotor rhythm of a methamphetamine treated rat. The rat was transferred from LD to DD after 3 months of methamphetamine treatment. Shadowed areas indicate the dark period.

relation between the activity onset and the light-off time became unstable and the phase angle difference between the circadian rhythm and the LD cycle changed. Several weeks later, the circadian rhythm showed signs of so-called relative coordination. Figure 5 illustrates an example of relative coordination produced by chronic treatment of methamphetamine. The phenomenon continued also in DD.

The changes in the phase angle difference of the circadian rhythm and the light cycle, and the occurrence of the relative coordination can be explained by one of the following three models. Firstly, the phenomena are observed when an intrinsic period of the circadian oscillation is altered and exceeds the limits of entrainment; secondly, when the intensity of an entraining light signal or the sensitivity of the circadian pacemaker to light is changed; and thirdly, although this model is still hypothetical, a relative coordination may take place when the internal coupling between two oscillators is altered. The latter possibility was proposed for the explanation of a similar phenomenon observed in the sleep-wakefulness cycle of human subjects who lived in a temporal isolation unit for a long period (34). The phenomenon was referred to as internal relative coordination.

Previously, amphetamine was reported to lengthen the period of

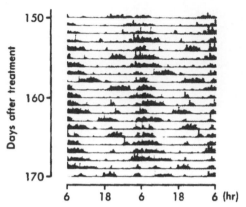

Fig. 6. Splitting of the circadian locomotor rhythm in a rat treated chronically with methamphetamine. The rhythm is double-plotted. Shadowed areas indicate the dark period.

the circadian rhythm (*30*). Although in our study the free-running period in DD did not differ significantly between the methamphetamine treated and control groups, the first model cannot be completely excluded. The rat retina contains dopamine neurons, whose activity is stimulated by the photoperiod (*10*). The dopamine content and turnover in the retina showed a day-night rhythm which was a direct result of the LD cycle (*20*). Therefore, it is possible that methamphetamine changes the sensitivity to light at the level of the retina. This possibility seems to be unlikely, however, because the changes of the circadian rhythm persisted even in DD. Figure 6 shows the locomotor rhythm 5 months after the beginning of methamphetamine treatment. In this stage, the evening component of the circadian rhythm was no longer entrained by the LD cycle and began to free-run. On the other hand, the morning component was still entrained to the LD cycle, though the phase relation to the LD cycle was unstable and changed periodically. The circadian rhythm was separated into two components, similar to splitting in LL. A Chi-square periodogram indicated that the period of the free-running component was around 30.0 hr in most rats.

Splitting and the partial entrainment of the circadian locomotor rhythm in methamphetamine treated rats can be readily accounted for by the two-oscillator pacemaking model mentioned above. When the internal coupling of two oscillators is weakened, the oscillator driving the evening component of the circadian rhythm begins to free-run in the 24-hr LD cycle, because the oscillator is unable to make a phase advance shift (the onset PRC lacking the phase advance area). A phase-advance is necessary for the entrainment to the 24-hr LD cycle if the intrinsic period of the oscillator is longer than 24 hr. The free-running period of the rat locomotor rhythm in our colony is almost exclusively longer than 24 hr. On the other hand, the other oscillator which is responsible for the morning component is capable of entraining to the 24-hr LD cycle, because it is able to phase advance in response to light impinging in the late subjective night.

Whatever the mechanism of splitting induced by methamphetamine may be, the findings suggest an involvement of central catecholaminergic mechanism in the circadian organization of locomotor activity in rats. The neurochemical background of the methamphetamine effects is not known. Methamphetamine releases dopamine and norepinephrine from the nerve terminal of catecholaminergic neurons. Prolonged administration of methamphetamine depletes the brain catecholamines, a part of which is explained by the destruction of the nerve fibers (26). However, the destruction of the catecholaminergic nerves cannot explain the effects of methamphetamine, because the relative coordination and splitting of the circadian rhythm disappeared immediately after replacing methamphetamine with normal drinking water. Since the drug effects were manifest only after a long latent period, altered sensitivities of the post-synaptic mechanism seem to be involved.

SUMMARY

The circadian pacemaker underlying the rest-activity cycle of rats is suggested to be comprised of at least two mutually coupled oscillators. The central cholinergic mechanism may be involved in the entrainment of the pacemaker to the LD cycle. On the other hand, the central catecholaminergic system seems to be related to the circadian organization, especially to the internal coupling of the two circadian oscillators.

REFERENCES

1 Aghajanian, G.K., Bloom, F.E., and Sheard, M.H. (1969). *Brain Res.* **13**, 266–273.

2 Albers, H.E., Ferris, C.F., Leeman, S.E., and Goldman, B.D. (1984). *Science* **223**, 833–835.

3 Borbély, A.A. (1982). In *Sleep: Clinical and Experimental Aspects, Current Topics in Neuroendocrinology*, ed. Ganten, D. and Pfaff, D., pp. 83–103. Berlin: Springer-Verlag.

4 Boulos, Z. and Terman, M. (1979). *J. Comp. Physiol.* **A134**, 75–83.

5 Boulos, Z. and Rusak, B. (1982). *J. Comp. Physiol.* **A146**, 411–417.

6 Brownstein, M.J., Kobayashi, R., Palkovits, M., and Saavedra, J.M. (1975). *J. Neurochem.* **24**, 35–38.

7 Cheung, P.W. and McCormack, C.E. (1983). *Am. J. Physiol.* **244**, R573–R576.

8 Earnet, D.J. and Turek, F.W. (1983). *Science* **219**, 77–79.

9 Ellis, G.B., McKlveen, R.E., and Turek, F.W. (1982). *Am. J. Physiol.* **242**, R44–R50.

10 Frucht, Y., Vidauri, J., and Melamed, E. (1982). *Brain Res.* **249**, 153–156.

11 Groos, G., Mason, R., and Meijer, J. (1983). *Fed. Proc.* **42**, 2790–2795.

12 Jouvet, M. (1969). *Science* **163**, 32–41.

13 Honma, K. and Hiroshige, T. (1979). *Brain Res.* **169**, 519–529.

14 Honma, K. and Hiroshige, T. (1979). *Brain Res.* **169**, 531–544.

15 Honma, K. (1985). In *Circadian Clocks and Zeitgebers*, ed. Hiroshige, T. and Honma, K., pp. 106–117. Sapporo: Hokkaido Univ. Press.

16 Honma, K., Honma, S., and Hiroshige, T. (1985). *Jpn. J. Physiol.* **35**, 645–660.

17 Ikeda, Y. and Chiba, Y. (1982). In *Toward Chronopharmacology*, ed. Takahashi, Y., Halberg, F., and Walker, C.A., pp. 3–10. New York: Pergamon Press.

18 Illnerová, H. and Vaněček, J. (1982). *J. Comp. Physiol.* **A145**, 539–548.

19 Krieger, D.T. (1975). *Neuroendocrinology* **17**, 62–74.

20 Melamed, E., Frucht, Y., Lemor, A., Uzzan, A., and Rosenthal, Y. (1984). *Brain Res.* **305**, 148–151.

21 Meyer, D.C. and Quay, W.B. (1976). *Endocrinology* **98**, 1160–1165.

22 Monti, J.M. (1982). *Life Sci.* **30**, 1145–1157.

23 Pittendrigh, C.S. (1960). *Cold Spring Harbor Symp. Quant. Biol.* **25**, 159–184.

24 Pittendrigh, C.S. (1974). In *The Neurosciences: Third Study Program*, ed. Schmitt, F.O. and Worden, F.G., pp. 437–548. Cambridge: MIT Press.

25 Pittendrigh, C.S. and Daan, S. (1976). *J. Comp. Physiol.* **106**, 333–355.

26 Ricaurte, G.A., Seiden, L.S., and Schuster, C.R. (1984). *Brain Res.* **303**, 359–364.

27 Riou, F., Cespuglio, R., and Jouvet, M. (1982). *Neuropeptides* **2**, 243–277.

28 Segal, M., Dudai, Y., and Amsterdam, A. (1978). *Brain Res.* **148**, 105–119.

29 Scapagnini, U., Moberg, G.P., Van Loon, G.R., and Ganong, W.F. (1970). *Eur. J. Pharmacol.* **11**, 266–268.

30 von Bodman, K. (1970). *Z. Vergl. Physiol.* **68**, 276–292.

31 Wirz-Justice, A. and Campbell, I.C. (1982). *Experientia* **38**, 1301–1309.

32 Zatz, M. and Brownstein, M. (1979). *Science* **203** 359–361.

33 Zarz M. and Herkenham, M.A. (1981). *Brain Res.* **212**, 234–238.

34 Zulley, J., Wever, R., and Aschoff, J. (1981). *Pflügers Arch.* **391**, 314–318.

9

EFFECTS OF CONTINUOUS INFUSIONS OF TRH AND GIF ON CIRCADIAN RHYTHM PARAMETERS OF SLEEP, AMBULATION, EATING, AND DRINKING IN RATS

YASURO TAKAHASHI, SETSUO USUI, SHIGEMITSU
EBIHARA, AND YOSHIKO HONDA

Department of Psychology, Tokyo Metropolitan Institute for Neurosciences, Fuchu 183, Japan

In recent years, it has been reported that many neuropeptides have a variety of actions regulating behavioral, endocrine and autonomic functions. Thyrotropin-releasing hormone (TRH) and growth hormone release inhibiting factor (GIF; somatostatin) are well known to be hypothalamic hypophysiotropic hormones and, in addition, to have some effects in modulating behaviors (*4, 11, 14, 17, 18*). These include an awakening effect of these neuropeptides (*3, 12*). Although endogenous sleep-promoting factors have attracted special attention recently, endogenous sleep-inhibiting or awakening factors also are important for understanding the humoral control of sleep. The behavioral effects of TRH and GIF reported previously, however, have been based upon short-term observations after a single administration. We have attempted to elucidate the effects of long-term continuous infusion of these peptides on chronobiological aspects of sleep and other spontaneous behaviors of rats.

I. TECHNICAL ASPECTS

We have developed an automatic apparatus for simultaneous and continuous measurements of sleep (slow wave sleep (SWS), and paradoxical sleep (PS)), ambulation, eating, and drinking, and a device for continuous micro-infusion of the peptides in freely moving rats. Male adult Sprague-Dawley rats (450–500 g) were chronically implanted with electrodes for recording the electroencephalogram (EEG), electromyogram (EMG), and body movements, and with a guide cannula in the lateral ventricle. Physiological saline and TRH or GIF saline solutions with two 10-fold different concentrations were infused at a rate of 1 μl/ hr for 7.5 days with Alzet osmotic minipumps (#2001). In regard to the biological activities of these peptides under this long-term infusion condition, TRH was quite stable whereas GIF was deactivated by 16% in a 1 nmol/μl solution. The intracerebroventricular (i.c.v.t.) infusion was done with an injector cannula attached *via* a silicone tubing to a minipump implanted subcutaneously. TRH and GIF occur naturally in the brain, blood, and some peripheral organs (*4, 11*). Since both the peptides are reported to have poor penetration of the blood-brain barrier and very short half-lives in blood (*4, 6, 11*), each peptide solution with higher concentration was subcutaneously (s.c.) infused in order to differentiate the effects on the central nervous system from the peripheral effects. For each peptide, all rats received the i.c.v.t. and s.c. infusions in random order, 1 to 2 weeks apart. The minipumps were implanted under penthrane anesthesia at 17:00. Thereafter, sleep, ambulation, eating, and drinking were recorded during the infusion period on a 12:12 hr light-dark cycle (light on 6:00 and off 18:00).

The hourly time series data of each behavior per animal were analyzed by the chi-square periodogram for periodicity (*13*), and by the cosinor method (*10*) for the mesor, amplitude and acrophase of 24-hr periodicity which was the most dominant and significant period under all the infusion conditions. The effects of the peptide infusions were analyzed by the group cosinor method (*10*). Serial section analysis (*1*) was applied to examine day-to-day changes in the cosinor parameters. The mesor and amplitude reflect the overall amount of each behavior and roughly its difference between the light and dark periods, respectively, under these experimental conditions.

II. EFFECT OF TRH INFUSION

In our preliminary study (15), 4 and 16 μg/hr i.c.v.t. infusions of TRH produced a marked awakening effect, but significantly dampened the circadian rhythmicity of sleep, particularly of PS. Thus, in the present study, 1 μg=2.8 nmol/hr and 0.1 μg/hr i.c.v.t. infusions of TRH were administered to 7 rats. A significant 24-hr periodicity was detected in all behavioral rhythms during the TRH infusions, but the rhythms of PS and eating were slightly dampened during the 1 μg/hr i.c.v.t. infusion (Fig. 1). As compared with the i.c.v.t. saline infusion, the i.c.v.t. TRH infusions significantly affected some cosinor parameters in a dose-dependent manner: decreased mesor and amplitude with advanced acrophase in the circadian rhythms of total sleep time (TST), SWS and PS; and increased mesor and amplitude with advanced acrophase in the ambulation rhythm (Fig. 2). When overall 24-hr profiles were compared between the icvt. infusions of saline and 1 μg/hr TRH, the sleep-sup-

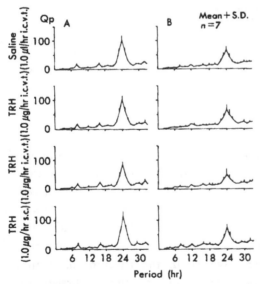

Fig. 1. Chi-square periodograms of SWS (A) and PS (B) based on 7.5-day hourly time series data from saline and TRH infusions. Dotted lines indicate 0.01 significance level.

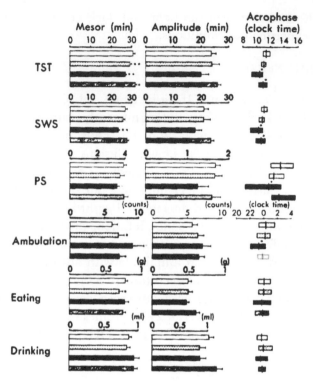

Fig. 2. Comparisons of mesor, amplitude, and acrophase of rhythms of sleep (upper), ambulation, eating, and drinking (lower) during TRH infusion in 7 rats. Mean+S.E.M. for mesor and amplitude, and mean with 95 % confidence limits for acrophase. P was determined by two-tailed paired *t*-test. ☐ saline 1.0 μl/hr i.c.v.t.; ■ TRH 1.0 μg/hr i.c.v.t.; ▨ TRH 0.1 μg/hr i.c.v.t.; ▨ TRH 1.0 μg/hr s.c. P: *vs.* saline. *p<0.05, **p<0.01.

pressing and ambulation-enhancing effects of TRH were not uniform throughout the day despite the presumed constant infusion rate (Figs. 3, 4). A marked enhancement of ambulation consistently occurred soon after dark onset (Fig. 4). The sleep-suppressing effect was most pronounced on day 1 of TRH infusion, whereas the ambulation-enhancing effect was almost uniform throughout the entire infusion period (Figs. 5, 6). The acrophase advance in the rhythms of sleep (both SWS and PS) and ambulation was greater during the first-half of the infusion period (Figs. 5, 6). Eating was suppressed only during the i.c.v.t. infusion of 0.1 μg/hr TRH (Fig. 2).

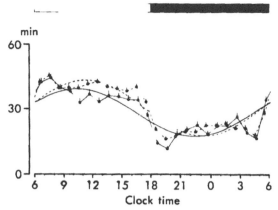

Fig. 3. Overall mean (+S.E.M.) 24-hr profiles of total sleep time for 7 rats during saline (O 1 μl/hr) and TRH (● 1 μg/hr) i.c.v.t. infusions. The best fitting cosine curve computed by the least square method is superimposed on each profile. Light-dark periods are indicated on the top.

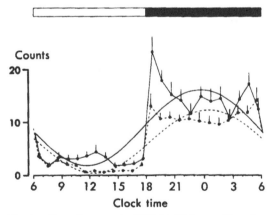

Fig. 4. Overall mean (+S.E.M.) 24-hr profiles of ambulation for 7 rats during saline (O 1 μl/hr) and TRH (● TRH 1 μg/hr) i.c.v.t. infusions. The best fitting cosine curve computed by the least square method is superimposed on each profile.

The TRH 1 μg/hr s.c. infusion produced a slight but significant acrophase advance only in the rhythms of TST and SWS in comparison to the saline infusion. The above-mentioned changes in the circadian rhythm parameters of sleep and ambulation were significantly greater during the i.c.v.t. infusion than during the s.c. infusion at the same rate

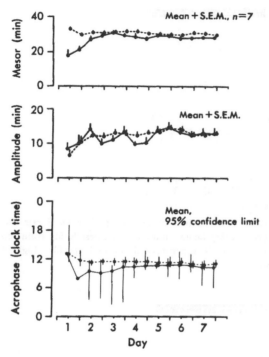

Fig. 5. Day-to-day variations of mesor, amplitude and acrophase of sleep rhythm obtained by progressive serial section analysis (period 24 hr; interval 24 hr; increment 12 hr) during saline (○ 1 µl/hr) and TRH (● 1 µg/hr) i.c.v.t. infusions in 7 rats.

of 1 µg/hr. This suggests that the effects of TRH are primarily attributable to its direct action on the central nervous system.

III. EFFECT OF GIF INFUSION

Eight rats received 5 treatments: the i.c.v.t. infusions of saline, GIF 1 nmol=1.6 µg/hr and 10 nmol/hr, and the s.c. infusion of saline and GIF 10 nmol/hr. These GIF infusions did not affect 24-hr periodicity in any behavioral rhythms. Unlike the TRH infusions, the day-to-day and inter-individual variability of all the cosinor parameters decreased during the GIF infusions.

The mesors of TST and SWS showed a slight dose-dependent decrease during the GIF i.c.v.t. infusions, whereas the mesor of PS tended to increase. The acrophases of the rhythms of TST, SWS and PS were

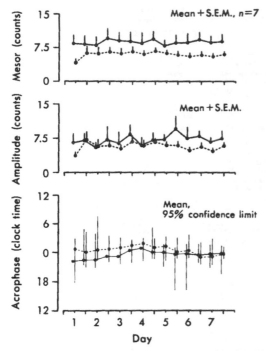

Fig. 6. Comparisons of mesor, amplitude, and acrophase of ambulation rhythm by progressive serial section analysis for saline (○ 1 μl/hr) and TRH (● 1 μg/hr) i.c.v.t. infusions in 7 rats.

advanced during the 10 nmol/hr i.c.v.t. infusion (Fig. 7). The day-to-day changes of these cosinor parameters were almost uniform during the 7.5-day infusion period (Fig. 8). Drinking was significantly suppressed at 10 nmol/hr, but eating was not (Fig. 7). An advanced acrophase in the rhythms of ambulation and eating was noted (Fig. 7).

There were no significant differences in the three cosinor parameters of any behavioral rhythm between the i.c.v.t. and s.c. saline infusions (Fig. 7). A significant acrophase advance was found in PS, ambulation and eating during the GIF 10 nmol/hr s.c. infusion in comparison to the saline s.c. infusion (Fig. 7).

Generally speaking, the changes in the circadian rhythm parameters during the GIF infusions were similar to, but slighter than those during the TRH infusions.

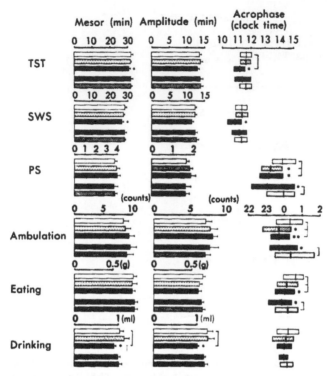

Fig. 7. Comparisons of mean mesor, amplitude, and acrophase of rhythms of sleep (upper), ambulation, eating, and drinking (lower) during GIF infusions in 8 rats. Mean +S.E.M. for mesor and amplitude, and mean with 95 % confidence limits for acrophase. ☐ saline 1.0 μl/hr i.c.v.t.; ■ GIF 10 nM/hr i.c.v.t.; ▨ GIF 1.0 nM/hr i.c.v.t.; ▨ GIF 10 nM/hr s.c.; ▨ saline 1.0 μl/hr s.c. *p<0.05, **p<0.01.

IV. CHRONOBIOLOGICAL CONSIDERATION

Data from continuous long-term recordings are necessary for the statistical analysis of circadian rhythms (*10, 13*). As is well known in chronopharmacology, the effects of many drugs vary with the time of their administration within a 24-hr period (*7*). Since during a repeated administration of a drug circadian effects are also present, data analysis is difficult. The long-term constant infusion technique used in the present study also has some problems. Even if a constant infusion rate is preserved, the sensitivity to the drug in the organism may vary as a function of time as has been reported for TRH (*5, 9*), and also other

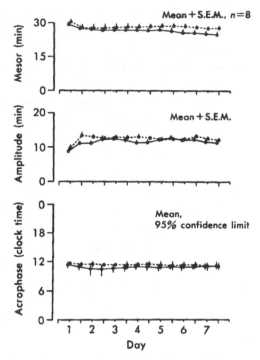

Fig. 8. Day-to-day variations of mesor, amplitude and acrophase of SWS rhythm obtained by progressive serial section analysis (period 24 hr; interval 24 hr; increment 12 hr) during saline (1 μl/hr) and GIF (10 nM/hr) i.c.v.t. infusions in 8 rats.

parameters, such as the metabolic clearance rate, *etc.*, may change with time.

Upon i.c.v.t. administration in rats, both TRH and GIF were reported to have a general arousal effect independent of their actions on the pituitary (*3*), and a suppressing effect on eating and drinking (*17*, *18*). On the other hand, differences between the mode of action of these neuropeptides on the neuronal activity of the brainstem reticular formation and the hippocampus in rats were reported (*5*). Enhanced locomotor activity is one of the well-known behavioral effects of TRH (*4*, *18*). The results obtained in our long-term continuous infusion study are in general agreement with those in the previous reports. However, PS was not suppressed by GIF in contrast to reports of a marked PS-inhibiting effect (*3*).

One of the most consistent effects of TRH and GIF in our study was acrophase advance in the circadian rhythms of behaviors, particularly of sleep and ambulation. There are two possible explanations for the finding that the phase position of circadian rhythm was advanced in the conditions with entrainment to a 24-hr light-dark cycle. One possibility is that these peptides may shorten the intrinsic circadian period of the behavioral rhythms. A circadian rhythm that exhibits a shortened period in free-running conditions will adopt an early phase position when entrained to a 24-hr environmental period (7). For example, estradiol which shortens the free-running period of locomotor activity of hamsters, caused a phase-advance under 24-hr entrained conditions (8, 16), whereas lithium and heavy water which lengthen the intrinsic circadian period, caused a phase-delay (7). There is ample evidence for the existence of a main oscillator in the suprachiasmatic nuclei (SCN) of rats controlling the circadian rhythms of various behavioral and physiological variables (7). Since both TRH and GIF are present in the SCN of rats (19), the peptides may play a role in the regulation of circadian rhythms. In addition, there are some indications for a relationship between the increased arousal level of animals and shortened circadian periods (16). The present results seem to be in accordance with this assumption, since TRH and GIF produced enhanced levels of wakefulness and ambulation in the present study.

If the intrinsic circadian periods are unchanged by TRH and GIF, an alternative explanation for the phase advance may lie in the enhanced sensitivity to zeitgebers, such as light (7). In animals with an intrinsic circadian rhythm period longer than 24 hr (*e.g.*, the rat), a higher sensitivity to light induces an earlier phase position of the circadian rhythm (7). Our observations revealed that in comparison to saline both TRH and GIF clearly enhanced ambulation and suppressed sleep at dark onset (Figs. 3, 4). The presence of TRH and GIF in the rat retina is suggestive of some roles of these peptides in retinal adaptation to light and dark conditions (2).

SUMMARY

The effects of 7.5-day continuous i.c.v.t. and s.c. infusions of TRH (0.1 and 1 μg/hr) and GIF (1 and 10 nmol/hr) on the circadian rhythm parameters of sleep, ambulation, eating, and drinking were studied in rats maintained on a 12:12 hr light-dark cycle. In a dose-dependent manner, TRH suppressed SW and PS and enhanced ambulation while advancing the acrophase of the circadian rhythms. GIF had similar but smaller effects than TRH. GIF did not suppress PS. While the effects of TRH were initially prominent and then decreased, those of GIF were almost uniform throughout the entire infusion period.

The acrophase advance in the circadian rhythms of sleep and ambulation may be due to shortened periods of intrinsic circadian rhythms and/or to an enhanced sensitivity to light and dark conditions.

Acknowledgments
The authors wish to thank Dr. S. Sawano, Toranomon Hospital, Tokyo for his radioimmunoassay of GIF. This work was supported by a Grant-in-Aid for Scientific Research (No. 57440059) and Grant-in-Aid for Special Project Research (No. 58106004) from the Ministry of Science, Education and Culture of Japan.

REFERENCES

1 Arbogast, B., Lubanovic, W., Halberg, F., Cornélissen, G., and Bingham, C. (1982). *Chronobiologia* 10 59–68.

2 Brecha, N.C. and Karten, H.J. (1983). In *Brain Peptides*, ed. Krieger, D.T., Brownstein, M.J., and Martin, J.B., pp. 437–462. New York: John Wiley & Sons.

3 Havlicek, V., Rezek, M., and Friesen, H. (1976). *Pharmacol. Biochem. Behav.* 4, 455–459.

4 Jackson, I.M.D. and Lechan, R.M. (1983). In *Brain Peptides*, ed. Krieger, D.T., Brownstein, M.J., and Martin, J.B., pp. 661–685. New York: John Wiley & Sons.

5 Korányi, L., Whitmoyer, D.I., and Sawyer, C.H. (1977). *Exp. Neurol.* 57, 807–816.

6 Meisenberg, G. and Simmons, W.H. (1983). *Life Sci.* 32, 2611–2623.

7 Moore-Ede, M.C., Sulzman, F.M., and Fuller, C.A. (1982). *The Clocks That Time Us*. Cambridge, London: Harvard Univ. Press.

8 Morin, L.P., Fitzgerald, K.M., and Zucker, I. (1977). *Science* 196, 305–307.

9 Nemeroff, C.B., Bissette, G., Martin, J.B., Brazeau, P., Vale, W., Kizer, J.S., and Prange, A.J., Jr. (1980). *Neuroendocrinology* 30, 193–199.

10 Nelson, W., Tong, Y.L., Lee, J.-K., and Halberg, F. (1979). *Chronobiologia* 6, 305–323.

11 Reichlin, S. (1983). In *Brain Peptides*, ed. Krieger, D.T., Brownstein, M.J., and Martin, J.B., pp. 711–752. New York: John Wiley & Sons.

12 Rezek, M., Havlicek, V., Hughes, K.R., and Friesen, H. (1976). *Pharmac. Biochem. Behav.* 5, 73–77.

13 Sokolove, P.G. and Bushell, W.N. (1978). *J. Theor. Biol.* 72, 131–160.

14 Stanton, T.L., Beckman, A.L., and Winokur, A. (1981). *Science* 214, 678–681.

15 Takahashi, Y., Usui, S., Ebihara, S., and Honad, Y. (1983). *Sleep Res.* 12, 136.

16 Turek, F.W. and Gwinner, E. (1982). In *Vertebrate Circadian Systems*, ed. Aschoff, J., Dann, S., and Groos, G.A., pp. 173–182. Berlin, Heidelberg: Springer-Verlag.
17 Vijayan, E. and McCann, S.M. (1977). *Endocrinology* 100, 1727–1730.
18 Vogel, R.A., Cooper, B.R., Barlow, T.S., Prange, A.J., Jr., Mueller, R.A., and Breese, G.R. (1979). *Exp. Ther.* 208, 161–168.
19 Zucker, I. and Carmichael, M.S. (1981). In *Neurosecretion and Brain Peptides*, ed. Martin, J.B., Reichlin, S., and Bick, K.L., pp. 459–473. New York: Raven Press.

10

CIRCADIAN RHYTHMS OF FEEDING AND METABOLISM: SUPRACHIASMATIC NUCLEUS AS A BIOLOGICAL CLOCK AND A SITE OF METABOLIC REGULATION

KATSUYA NAGAI, HIDEKI YAMAMOTO, AND HACHIRO NAKAGAWA

Division of Protein Metabolism, Institute for Protein Research, Osaka University, Osaka 565, Japan

During our study to elucidate the mechanism of generation of the daily rhythm of gluconeogenesis, we found that the phase of the rhythm of gluconeogenic enzyme activities in the liver and kidney of rats depended on the time of food intake rather than on the light-dark cycle. Thus we started to investigate the mechanism of generation of daily rhythms in feeding behavior and metabolism of rats. In this paper we describe our findings concerning the mechanisms of generation of the circadian rhythm of feeding behavior, temporal memory formation, and metabolism. We address in particular the problem of how mammals maintain glucose homeostasis under a daily cycle of sleep-wakefulness.

I. DAILY RHYTHMS IN ACTIVITIES OF PHOSPHOENOLPYRUVATE CARBOXYKINASE

In the study of the regulation of gluconeogenesis, we found daily rhythms in activities of phosphoenolpyruvate carboxykinase (PEPCK), a key enzyme of gluconeogenesis, in the liver and kidney of rats. Under a 12-hr light and 12-hr dark cycle, the activity of liver PEPCK was

low at the beginning of the light period and high at the beginning of
the dark period, while the activity of kidney PEPCK was low at the
beginning of the dark period and high during the later part of the dark
period (*16*, *19*). Hence the two rhythms showed a phase-difference of
about 6 to 12 hr. In the experiment using radioactive precursors of
gluconeogenesis *in vivo*, we found that changes in the rate of gluconeogenesis in the liver and kidney paralleled the changes in activities of
PEPCK in these respective organs (*3*, *4*). Treatments increasing the
sympathetic neural activity such as administrations of adrenaline, noradrenaline, and atropine, as well as vagotomy elevated liver enzyme
activity and lowered kidney enzyme activity. On the other hand, treatments increasing the parasympathetic neural activity such as administration of carbamylcholine and bilateral lesions of the ventromedial
hypothalamus (VMH), lowered liver enzyme activity and elevated
kidney enzyme activity (*10*, *15*). These findings suggest that the daily
rhythms of PEPCK activity in the liver and kidney are generated
by the daily alternation in activity of the autonomic nervous systems.
Liver enzyme activity is increased by the activation of the sympathetic
nervous system and is decreased by the activation of the parasympathetic nervous system, whereas kidney enzyme activity changes in the
opposite direction. The daily rhythms of gluconeogenesis in the liver
and kidney may also be generated by these mechanisms.

II. CIRCADIAN RHYTHMS OF FEEDING BEHAVIOR

1). *Central Mechanism of the Rhythm*
Since the phases of daily rhythms of PEPCK activities in the liver and
kidney depend on the feeding time rather than on the light-dark cycle,
we investigated the mechanism of generation of the daily rhythm of
feeding behavior in rats. When kept under a 12-hr light and 12-hr dark
cycle, rats eat about 80–90% of their total daily food intake during the
12-hr dark period. After orbital enucleation, they usually show a free-
running circadian rhythm of food intake with a period close to 24 hr.
This demonstrates that the daily rhythm of feeding behavior of rats is
a circadian one. Since Stephan and Zucker (*28*) and Moore and Eichler
(*6*) independently reported in 1972 that bilateral lesions of the suprachiasmatic nucleus (SCN) eliminated circadian rhythms of drinking

behavior, locomotive activity, and blood corticosterone concentration in rats, many circadian rhythms including the sleep-wakefulness cycle have been shown to disappear after the bilateral lesions of the SCN (24). We found that bilateral electrolytic and immunologic (using anti-rat SCN rabbit antibody) lesions of the SCN completely abolished the circadian rhythm of feeding behavior of rats (11, 13). This finding suggests that the SCN is an endogenous oscillator of the circadian feeding behavior of this animal. With regard to underlying mechanisms, Balagura and Devenport (1) suggested that the VMH is another candidate for the biological clock, because bilateral VMH lesions disrupted the circadian feeding rhythm in rats. However, in our own experiments VMH lesions did not completely eliminate the circadian rhythm, since the percentage of food intake during the 12-hr light period was not about 50% of total daily food intake, but fluctuated between 3 to 40% (20). Furthermore, we obtained evidence for the existence of neural connections between the SCN and the lateral hypothalamus (LH; the "feeding center"), and between the SCN and the VMH (the "satiety center") (20, 21). These findings suggest that the circadian feeding behavior of rats is generated by a time signal which originates in the SCN and is transmitted to other nuclei such as the VMH and the LH. Consequently, the disturbance of the circadian feeding rhythm after VMH-lesions may be due to the disruption of fiber connections from the SCN to the LH and VMH. The data of Rietveld et al. (23) support this interpretation. In addition to the circadian feeding rhythm, the daily activity rhythms of pineal serotonin N-acetyltransferase (SNAT), a rate-limiting enzyme of melatonin synthesis, and of liver PEPCK were retained in rats after bilateral lesions of the VMH (10, 18), although after the VMH-lesions the activity levels and amplitudes of the rhythms were altered. Similar to the reports of Moore and Klein (7) that bilateral SCN lesions eliminated the circadian rhythm of pineal SNAT activity, we observed that SCN lesions abolished the daily rhythms of PEPCK activity in the liver and kidney of rats (20). Moreover, in rats kept on a restricted feeding schedule (3-hr food access in the light period for 25 days) the phase of the daily rhythms of PEPCK activity in the liver and kidney was shifted, but not that of pineal SNAT activity (18). These findings suggest that the circadian rhythms of feeding behavior and pineal SNAT activity are both generated by time signals originating in the SCN, while the daily rhythms in PEPCK activity in liver and kidney depend on and are synchronized by the circadian feeding rhythm.

2). Effects of Various Agents Infused into the SCN

In order to study the central mechanism of the circadian feeding behavior, insulin and other agents were infused by an Alzet–osmotic minipump for one week into the SCN of rats maintained under a 12-hr light and 12-hr dark cycle. Insulin infusion reversibly disturbed the circadian rhythm of feeding behavior; compared with saline infusion, insulin did not affect the total daily food intake, but increased the percentage eaten in the 12-hr light period (12). Since histaminergic nerve fibers were recently shown to reach the SCN, we also examined the effect of histamine infusion into the SCN. Similar to insulin, histamine did not change the total daily food intake, but increased the percentage eaten in the light period (N. Itoi et al., unpublished observation). By contrast, the infusion of another peptide hormone modified at the N-terminal, glutaryl-CCK-8, decreased the total daily food intake by reducing it only in the dark period, thereby increasing its percentage in the light period (T. Mori et al., unpublished observation). In summary, infusion of insulin, histamine, and the calciumionophore A23187 (20) disturbed the circadian rhythm of feeding behavior without affecting the total daily food intake, while the infusions of glutaryl-CCK-8, insulin-like growth factor I (S. Takagi et al., unpublished observation) and mazindol (14), an anorectic drug, reduced total daily food intake by decreasing the food intake only in the dark period (Table I). These results suggest that insulin, histamine, and calcium are involved in the generation of the circadian feeding behavior by acting on the SCN, while glutaryl-CCK-8, insulin-like growth factor I and mazindol elicit hypophagia by affecting neurons in the VMH and LH.

TABLE I
Effects of Various Agents on Feeding Behavior of Rats

Total daily food intake (24 hr)	→	↓
Food intake (in 12-hr light period)	↑	→
Food intake (in 12-hr dark period)	↓	↓
	Insulin	CCK-8
	Histamine	IGF-I
	Ca-ionophore (A23187)	Mazindol

III. TEMPORAL MEMORY FORMATION AND THE SCN

Rats maintained under a 12-hr light and 12-hr dark schedule and *ad libitum* feeding usually drink about 80–90% of their daily water intake during the dark period. However, during a 3-week restricted feeding schedule with food access from 14:00 to 17:00, 70 to 80% of the daily water intake occurred during the restricted feeding time, even though nocturnal drinking was still observed. After termination of the restricted feeding schedule, the drinking pattern immediately became nocturnal again, although a small portion of drinking persisted at the time of the previous restricted feeding for at least 10 days (*8*). Since prandial drinking represents 80 to 90% of total drinking, it is proposed that the persistence of drinking during the previous feeding time is due to temporal memory. The daily rhythm of water intake was abolished after bilateral lesions of the SCN. Under a restricted feeding schedule these SCN-lesioned animals showed only prandial drinking with water intake preceding feeding by no more than 2 hr (*8*). After cessation of the restricted feeding schedule, the drinking pattern immediately became acircadian, and therefore no evidence for a temporal memory could be detected.

We also examined the effect of blinding on intact rats at the time of transition from the 3-week restricted feeding schedule to the free feeding schedule. In this case the nocturnal drinking behavior under *ad libitum* feeding free-ran, and the part of drinking related to the prior feeding period free-ran with the same period (Fig. 1A) (*8*). In a further experiment, we observed the feeding behavior of rats which were first made blind, then subjected to the restricted feeding schedule for 5 months, and finally allowed free access to food. These animals showed a feeding behavior at the previous restricted feeding time until the day 8 after the end of restricted feeding, and in addition, a free-running nocturnal feeding behavior (Fig. 1B).

These findings suggest that 1) rats can memorize the time of feeding; 2) the SCN participates in this memory formation; 3) under condition in which the SCN is synchronized by the light-dark cycle, animals memorize the time using the biological clock in the SCN, and under condition in which the neural activity of the SCN is free-running

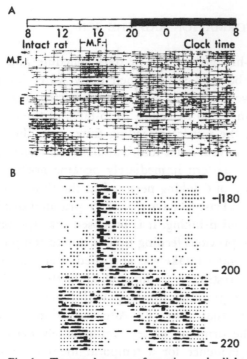

Fig. 1.   Temporal memory formation under light and dark and blind conditions. A: meal feeding (restricted feeding) was limited to 14:00- 17:00 for 3 weeks under 12-hr light and 12-hr dark conditions. M.F., start of restricted meal feeding; E, bilateral orbital enucleation. B: orbital enucleation was performed on day 16, restricted meal feeding from 14:00–17:00 was started on day 55, and was terminated on day 199 (S. Takagi *et al.* unpublished observation).

and not synchronized by the light-dark cycle animals memorize the time using a clock mechanism of the hourglass type. Whether the SCN is also involved in the latter mechanism is presently under investigation.

IV.   SCN AS A GLUCOREGULATORY CENTER

Since the central nervous system of mammals usually requires glucose as the only energy source, blood glucose level must be maintained. Although mammals exhibit a daily rhythm in feeding behavior, the energy expenditure of the brain is high throughout 24 hr. Thus, an extremely elaborate mechanism regulating blood glucose level must exist. Recently, we obtained evidence that the SCN is a glucoregulatory site which acts as the highest center in the regulation of glucose homeostasis. We examined the effect of insulin injection into the SCN on blood glucose concentration in the light and dark period in rats anesthetized by pentobarbital. While insulin injection into the SCN lowered the blood glucose level in the light period, it elevated the level in the dark period within 2 min after injection (9). The blood insulin level changed in parallel with the blood glucose level both in the light and dark period (9). Among glycogen metabolizing enzymes, the activity of liver glycogen phosphorylase *a* decreased, and that of liver glycogen synthase I increased within 2 min after insulin injection in the light period. On the other hand, the activity of liver glycogen synthase I tended to decrease within 2 min after the insulin injection in the dark period, as compared to activity after saline injection (9). Bilateral lesions of the SCN completely abolished these changes both in the light and dark period (9). These findings suggest that by acting on the SCN, insulin administered in the light period reduces glycogen phosphorylase *a* activity, increases glycogen synthase I activity in the liver *via* the hepatic branch of the autonomic nervous system, and reduces the blood glucose level, while after administration in the dark period, it decreases glycogen synthase I activity in the liver *via* the hepatic branch of the autonomic nervous system, and increases blood glucose concentration.

We also examined the effect of 2-deoxy-D-glucose (2DG) on glucose metabolism. Since it had been reported that peripheral injection of 2DG elicited a time-dependent hyperphagic response in rats (5), we examined whether in the unanesthetized rat, injection of 2DG into the lateral cerebral ventricle *via* a chronic cannula caused a time-dependent hyperglycemia. Saline injection did not cause any appreciable change in blood glucose level in either the light or dark period. However, 2DG caused greater hyperglycemia in the light period than in the dark period (17). Moreover, injection of D-mannitol and D-glucose elicited hyperglycemia in the light and dark period, respectively (17). In these experiments the plasma insulin concentration did not show any significant change after injection of these sugars or saline. In SCN-lesioned rats, the plasma insulin level was higher than in sham-operated control rats both before and after 2DG injection, and the hyperglycemic response to the 2DG injection was not seen (30). Similar results were obtained after intracranial injections of D-mannitol and D-glucose (32);

that is, bilateral lesions of the SCN completely eliminated the time-dependent hyperglycemic action of these sugars.

In these experiments we used rats that had been starved for 2 hr before sugar injections, and the SCN-lesioned rats had a higher pre-injection blood insulin level than the sham-operated controls. Thus it was possible that insulin secretion from the pancreas was enhanced in the SCN-lesioned rats. We investigated this possibility by administering an oral and intravenous glucose tolerance test. We found that the SCN-lesioned rats showed enhanced secretion of insulin and higher glucose tolerance in response to glucose administration (*31*).

These findings suggest that there are glucose receptive neurons in the SCN and that these neurons suppress the insulin secretion in the pancreas when they are active. Intracranial injection of 2DG elicits hyperglucagonemia, but there was no time-dependency in this hyperglucagonemic response (*34*). In SCN-lesioned rats plasma glucagon levels before and after injection of either 2DG or saline were lower than in sham-operated rats, and no hyperglucagonemia was induced by 2DG in either the light or dark period (*34*). These findings suggest that the SCN has a stimulatory action on glucagon secretion. We also found that 2DG injection induced hyperphagia and increase in the plasma free fatty acid concentration only in the light period, and that bilateral lesions of the SCN completely abolished these time-dependent responses (*33*).

Table II summarizes effects of brain lesions and adrenalectomy on the response to 2DG administration which were reported by us and others. None of the operations (*i.e.*, bilateral lesions of the VMH and the LH; combination of bilateral lesions of the VMH and the LH;

TABLE II
Effects of Operations on Responses to Intracranial Injection of 2DG

	Hyperglycemia	Lipolysis	Hyperphagia	Hyperglucagonemia
SCN lesions	−	−	−	−
VMH lesions	+	±	− or +	?
LH lesions	+	?	− or +	?
VMH lesions + LH lesions	+	?	?	?
Adrenalectomy	−	+	?	?

− disappeared; + appeared; ± partly disappeared; ? unknown.

adrenalectomy) except for bilateral SCN lesions eliminated all these responses to 2DG administration (5, 29, 30, 33, 34). The daily rhythms of plasma concentrations of glucose, insulin, glucagon, and free fatty acid were observed in fed sham-operated control rats. However, bilateral lesions of the SCN abolished all these daily rhythms and set the level of glucose and free fatty acid at the minimum values of controls, the level of insulin at the median value of controls, and the level of glucagon at about half the minimum of controls (35).

Until recently it was assumed that the VMH and the LH were the highest central control centers for blood glucose (22, 26). However, our own findings and those of others which include electrophysiological data suggest that the SCN sends via neural connections mainly stimulatory signals to the VMH, and inhibitory signals to the LH (20–22). In view of these findings and our own experimental results described above, it is hypothesized that the SCN monitors the blood concentrations of glucose and insulin. By increasing its neural activity, the SCN elevates glycogenolytic enzyme activity in the liver, inhibits insulin secretion, facilitates glucagon secretion, and thereby participates in control of the blood glucose concentration. Results of experiments using radioactive 2DG (27) and electrophysiological techniques (2) have suggested that the neural activity of rat SCN is high in the light period and low in the dark period. In view of the phase-relation between the daily rhythms of SCN activity and blood level of glucose, insulin, glucagon, and free fatty acid in fed rats, it is hypothesized that in the light period the SCN inhibits insulin secretion, stimulates glucagon secretion (blood insulin and glucagon concentrations were lower and higher in the light period, respectively), reduces glucose utilization and increases lipid utilization. Conversely, in the dark period the SCN elevates insulin secretion, lowers glucagon secretion (blood insulin and glucagon levels were higher and lower in the dark period, respectively), increases glucose utilization and decreases lipid utilization.

Since the SCN functions as a biological clock, the question arises whether the neurons that have a clock function have also a glucoregulatory function. Our preliminary results suggest that this is not the case. We monitored the free-running circadian rhythm of locomotor activity of individual blind rats, injected 2DG into the lateral cerebral ventricle in the subjective light period, and determined changes in blood glucose concentration. The hyperglycemic action of 2DG was observed until 3 weeks after orbital enucleation, however, it disappeared between 4 and 6 weeks after the blinding, and reappeared 8 weeks after blinding. Thus, we could dissociate the disappearance of the circadian rhythm and that of the hyperglycemic action of 2DG, since, while the clock mechanism was functional, the glucoregulatory response had disappeared. Recent electrophysiological findings suggested that 10 days after blinding, the neural activity in the ventrolateral part of the SCN decreased, but that neural activity did not change in the dorsomedial part (25). This finding is consistent with the conclusion that functions of clock and glucoregulation reside in different neurons of the SCN.

In conclusion, we speculate that the following central mechanism prevails between the circadian feeding behavior and metabolism (Fig. 2); 1) in the SCN there are at least 2 types of neurons, one type participating in the generation of circadian rhythms, and the other in the regulation of glucose homeostasis. 2) The neurons with biological clock function elicit the circadian rhythm of feeding through the VMH and the LH. 3) Neurons participating in glucose regulation receive timing

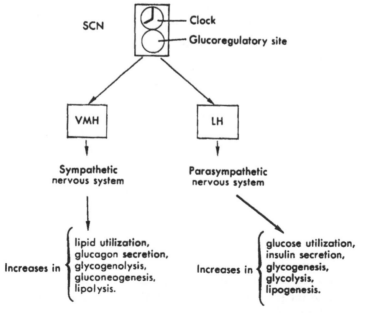

Fig. 2.   Hypothetical regulatory mechanism of the circadian rhythm of metabolism.

signals from the clock neurons and regulate glucose and other aspects of metabolism *via* the VMH, the LH and the autonomic nervous systems. 4) Enhanced activity of glucoregulatory neurons activates the neuronal activity in the VMH, which in turn enhances the activity of the sympathetic nervous system, which in turn inhibits insulin secretion and stimulates glucagon secretion. By the direct effects of these hormones and also *via* neural effect on the liver and adipose tissue, glycogen degradation, gluconeogenesis, and lipolysis are increased. Conversely, if the neuronal activity in the SCN decreases, activity in the LH and consequently in the parasympathetic nervous system increases. These changes stimulate insulin secretion, inhibit glucagon secretion, and *via* these hormones as well as *via* direct effects on the liver and adipose tissue, enhance glycogen synthesis, glycolysis, and lipogenesis. In this way the SCN may have a glucoregulatory function. To confirm this hypothesis further investigations are needed. Furthermore, on the basis of our findings, we speculate that the SCN may contain other control centers of physiological phenomena which are regulated by the autonomic nervous systems. This speculation also requires further examination.

SUMMARY

The mechanisms of the daily rhythms of feeding and metabolism in rats were examined. Bilateral lesions of the SCN completely eliminated daily rhythms of feeding and activities of PEPCK, a gluconeogenic enzyme, in the liver and kidney. In contrast, bilateral lesions of the VMH did not abolish the daily rhythms of feeding and liver PEPCK activity. Evidence was obtained for neural connections between the SCN and the VMH and between the SCN and the LH. These results suggest that daily rhythms of feeding behavior and PEPCK activities are generated by time signals from the SCN, which are transmitted to the VMH and the LH. Insulin, histamine, and calcium may be involved in this mechanism. Temporal memory is suggested to be formed by a biological clock in the SCN of rats under a light-dark cycle, but by a clock mechanism of the hourglass type in blind rats. Injection of insulin into the SCN caused hypoglycemia in the light period and hyperglycemia in the dark period. Intracranial injection of 2DG elicited time-dependent hyperglycemia, lipolysis and hyperphagia, and time-independent hyperglucagonemia. SCN-lesions induced insulin-hypersecretion and glucagonhyposecretion, and eliminated all the responses to insulin and 2DG. These findings suggest that the SCN has a glucoregulatory function besides functioning as a clock.

Acknowledgments

The authors wish to express their thank to Dr. Takashi Nishio, Dr. Tsutomu Mori, Dr. Shuji Inoue, Dr. Shunji Takagi, Mrs. Nobuko Itoi, and other investigators at the Division of Protein Metabolism, Institute for Protein Research, Osaka University, for their help and cooperation. The authors also would like to express their deep gratitude to Mr. T. Taniguchi, Founder of Taniguchi Foundation, for providing the opportunity to present this work at the 8th International Symposium on "Humoral Control of Sleep and Its Evolution."

REFERENCES

1 Balagula, S. and Devenport, L.D. (1970). *J. Comp. Physiol. Psychol.* **71**, 357–364.
2 Inouye, S.T. and Kawamura, H. (1979). *Proc. Natl. Acad. Sci. U.S.* **76**, 5962–5966.
3 Kida, K., Nishio, T., Nagai, K., Matsuda, H., and Nakagawa, H. (1982). *J. Biochem.* **91**, 755–760.
4 Kida, K., Nishio, T., Yokozawa, T., Nagai, K., Matsuda, H., and Nakagawa, H. (1980). *J. Biochem.* **88**, 1009–1013.
5 Le Magnen, J. (1983). *Physiol. Rev.* **63**, 314–386.
6 Moore, R.Y. and Eichler, V.B. (1972). *Brain Res.* **42**, 201–206.
7 Moore, R.Y. and Klein, D.C. (1974). *Brain Res.* **71**, 17–33.
8 Mori, T., Nagai, K., and Nakagawa, H. (1983). *Physiol. Behav.* **30**, 259–265.
9 Mori, T., Nagai, K., and Nakagawa, H. (1985). *Am. J. Physiol.* **249**, R23–R30.
10 Nagai, K., Inoue, S., and Nakagawa, H. (1983). *Am. J. Physiol.* E14–E18.
11 Nagai, K., Mori, T., and Nakagawa, H. (1982). *Biomed. Res.* **3**, 294–302.
12 Nagai, K., Mori, T., Nishio, T., and Nakagawa, H. (1982). *Biomed. Res.* **3**, 175–180.
13 Nagai, K., Nishio, T., Nakagawa, H., Nakamura, S., and Fukuda, Y. (1978). *Brain Res.* **142**, 384–389.
14 Nagai, K., Mori, T., Ookura, M., Tsujimoto, H., Takagi, S., and Nakagawa, H. (1985). *J. Obesity Weight Regul.* **4**, 33–49.
15 Nagai, K., Suda, M., and Nakagawa, H. (1973). *J. Biochem.* **74**, 863–871.
16 Nagai, K., Suda, M., Yamagishi, O., Toyama, Y., and Nakagawa, H. (1975). *J. Biochem.* **77**, 1249–1254.
17 Nagai, K., Yamamoto, H., and Nakagawa, H. (1982). *Biomed. Res.* **3**, 288–293.
18 Nagai, K., Yamazaki, K., Tsujimoto, H., Inoue, S., and Nakagawa, H. (1984). *Life Sci.* **35**, 769–774.

19 Nakagawa, H. and Nagai, K. (1971). *J. Biochem.* **69**, 923–934.
20 Nakagawa, H., Nagai, K., Kida, K., and Nishio, T. (1979). In *Biological Rhythms and Their Central Mechanism*, ed. Suda, M., Hayaishi, O., and Nakagawa, H., pp. 283–294. Amsterdam: Elsevier/North-Holland Biomedical Press.
21 Nishio, T., Shiosaka, S., Nakagawa, H., Sakumoto, T., and Satoh, K. (1979). *Physiol. Behav.* **23**, 763–769.
22 Oomura, Y., Ono, Y., Nishino, H., Kita, H., Shimizu, N., Ishizuka, S., and Sasaki, K. (1979). In *Biological Rhythms and Their Central Mechanism*, ed. Suda, M., Hayaishi, O., and Nakagawa, H., pp. 295–308. Amsterdam: Elsevier/North-Holland Biomedical Press.
23 Rietveld, W.J., Ten Hoor, F., Kooij, M., and Flory, W. (1978). *Physiol. Behav.* **21**, 615–622.
24 Rusak, B. and Zucker, I. (1979). *Physiol. Rev.* **59**, 449–526.
25 Shibata, S., Liou, S.Y., Ueki, S., and Oomura, Y. (1984). *Brain Res.* **302**, 75–81.
26 Shimazu, T. (1983). In *Advances in Metabolic Disorders*, ed. Szabo, A.J., vol. 10, pp. 355–384. New York, London: Academic Press.
27 Schwartz, W.J. and Gainer, H. (1977). *Science* **197**, 1089–1091.
28 Stephan, F.K. and Zucker, I. (1972). *Proc. Natl. Acad. Sci. U.S.* **69**, 1583–1586.
29 Yamamoto, H., Nagai, K., and Nakagawa, H. (1983). *Biomed. Res.* **4**, 505–514.
30 Yamamoto, H., Nagai, K., and Nakagawa, H. (1984). *Biomed. Res.* **5**, 55–60.
31 Yamamoto, H., Nagai, K., and Nakagawa, H. (1984). *Biomed. Res.* **5**, 47–54.
32 Yamamoto, H., Nagai, K., and Nakagawa, H. (1984). *Brain Res.* **304**, 237–241.
33 Yamamoto, H., Nagai, K., and Nakagawa, H. (1984). *Physiol. Behav.* **32**, 1017–1020.
34 Yamamoto, H., Nagai, K., and Nakagawa, H. (1985). *Endocrinology* **117**, 468–473.
35 Yamamoto, H., Nagai, K., and Nakagawa, H. (submitted to *Am. J. Physiol.*)

11

NEURAL CONTROL OF CIRCADIAN REST-ACTIVITY RHYTHM IN CRICKETS

KENJI TOMIOKA

Environmental Biology Laboratory, Biological Institute, Yamaguchi University, Yamaguchi 753, Japan

The majority of animals show daily rest-activity rhythms which have been demonstrated to be under circadian control. The circadian rhythm is generally thought to be generated in the central nervous system (CNS). Two approaches are demanded for the search for physiological mechanisms underlying the circadian rest-activity rhythm. One is the anatomical localization of the components of the circadian system: for example, the driving oscillator (pacemaker), the photoreceptor, and the driven system. The other is to elucidate how these components are integrated to form a regulatory system.

Insects are suitable material for such research, since the organization of the neural and endocrine systems that are ultimately responsible for the manifestation of the organized circadian rhythm is quite simple in comparison with that of vertebrates. Therefore, insects have been studied extensively in this particular field. The studies have been concentrated mostly on the rhythm of rest-activity represented by locomotion, stridulation or flight. For some insects the circadian pacemaker is reported to be located, though putatively, in the brain (Table I).

The locomotor rhythm in *Drosophila melanogaster* and *Acheta domes-*

TABLE I
Putative Site of Circadian Pacemaker Driving Rest-activity Rhythms

Organism	Function	Pacemaker site	Reference
Silkmoth			
Antheraea perni	Flight	Cerebral lobe	25
Hyalophora cecropia	Flight	Cerebral lobe	25
Fruitfly			
Drosophila melanogaster	Locomotion	Brain	4
House fly			
Musca domestica	Locomotion	Cerebral lobe	5
Mosquito			
Culex pipiens pallens	Flight	Cerebral lobe	2
Cockroach			
Leucophaea maderae	Locomotion	Optic lobe	10–15
Periplaneta americana	Locomotion	Optic lobe	10
Cricket			
Acheta domesticus	Locomotion	Brain	3
Teleogryllus commodus	Locomotion	Optic lobe	20
	Stridulation	Optic lobe	8, 20
Gryllus bimaculatus	Locomotion	Optic lobe (lamina + medulla)	23

ticus has been shown in tissue transplantation experiments to be controlled by brain-centered pacemakers (*3, 4*). In these cases there is apparently a humoral link somewhere in the output pathway of the pacemaker. In other insects, putative pacemaker sites have been localized in the brain using surgical lesions (*2, 5, 8, 10, 11, 20, 23, 25*). However, since none of these tissues has been demonstrated to contain a self-sustained oscillator, the physiological mechanism of interaction between the pacemaker and the driven system has been studied only to a somewhat limited extent.

These circumstances prompted us to make efforts to clarify the circadian system that controls the rest-activity rhythm using the male cricket *Gryllus bimaculatus*. The systematic investigation yielded almost certain evidence for the self-oscillating ability of the part of optic lobe containing the lamina and medulla. We also have been able to characterize to some extent the neural output of the lamina plus medulla region. This paper reviews these data and compares them with those obtained in other insects.

I. CIRCADIAN REST-ACTIVITY RHYTHM (21)

The research on the cricket rhythm was carried out mainly by means of measuring locomotor activity as an index of the rest-activity rhythm. The male cricket becomes adult after 8 successive molts. During the course of post-embryonic development, crickets show a marked reversal in the phasing of their circadian locomotor rhythm; the activity peak which is diurnal in the nymphs (nymphal rhythm=NR) becomes nocturnal (adult rhythm=AR) 3 to 5 days after the imaginal molt when signs of sexual maturation begin to show up (Fig. 1A). Although both NR and AR free-ran in constant darkness (DD) or constant light (LL,

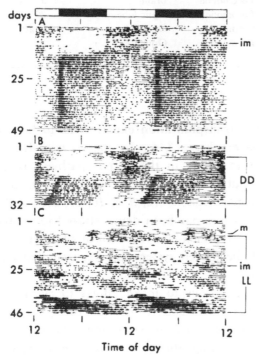

Fig. 1. Locomotor rhythm of male cricket under a light-dark cycle (LD) (A), DD (B), and LL (C) (21). The crickets were 8th (A, B) or 7th (C) instar on day 1. Tracings were double-plotted to facilitate the visual evaluation of changes of the activity peak. White and black bars on top indicate lighting regimen: ☐ light; ■ darkness. im, imaginal molt; m, 7th molt.

8–12 lux), they differed in phase, in the free-running period (τ) and in the waveform (Fig. 1B, C). τ_{DD} was significantly longer in NR (24.33 hr) than in AR (23.91 hr). AR was characterized by a sharp activity peak in each cycle, which NR lacked. To account for these differences, two explanations may be advanced; one assumes that NR and AR are separate oscillations, the other that both are coupled to different phase points of a single oscillation.

II. PHOTORECEPTOR FOR ENTRAINMENT

Surgical transection of the optic nerves abolishes entrainment by light; both NR and AR became free-running under a light-dark cycle (LD) just as in the intact animals maintained in DD (Fig. 2A). This fact indicates that, as in cockroaches (*Periplaneta americana* and *Leucophaea maderae*) (9, 17) and other cricket species (*Teleogryllus commodus*) (20), the

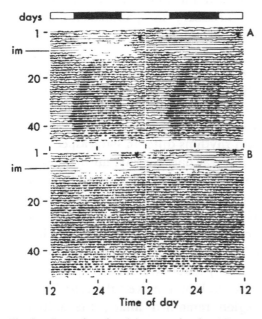

Fig. 2. Examples of activity records after bilateral optic nerve transection (A) or *optic lobe* removal (B) (23). Arrows indicate the time of operation. The lighting regimen was LD 12:12 throughout the experiments. In A, the adult rhythm became entrained roughly to the onset of darkness, probably due to partial regeneration of the optic nerves.

compound eyes of *G. bimaculatus* are the necessary source of photorecep-
tive information for entrainment of the circadian locomotor rhythm
(*23*). The reentrainment by the LD cycle after regeneration of the optic
nerves further supports this interpretation. It appears that the ocelli are
neither necessary nor sufficient for entrainment.

III. LOCUS OF THE DRIVING OSCILLATOR

The optic lobe is the site of the circadian pacemaker controlling the
locomotor rhythm in cockroaches. Nishiitsutsuji-Uwo and Pittendrigh
(*10*) were the first to suggest that bilateral ablation of the optic lobes or
sectioning of the optic tracts causes persistent arrhythmicity. This hypothesis has been supported and extended by subsequent lesion experiments (*11–15, 18, 19*). The optic lobes of the cockroach have three
distinct neuropile regions: lamina (the most distal part), medulla, and
lobula (the most proximal part). Since surgical or electrolytic lesions
on the ventral half of the lobe near the lobula frequently abolish the
rhythm (*11, 18, 19*), the cells necessary for pacemaker function are
suggested to have their soma and/or crucial process in the region of the
lobula (*11*).

In adult crickets (*T. commodus*), optic lobes also have been reported
to be indispensable for activity rhythms (*8, 20*). We also made surgical
lesions around the optic lobes in *G. bimaculatus* to study the following two
questions. First, does the hypothesis based only on AR also apply to
NR? As mentioned before, both rhythms differ in some aspects. Second,
is the lobula region the most integral part for rhythm manifestation in
the cricket? The optic lobes of crickets and cockroaches differ in structure; the swollen part near the retina contains only two neuropiles, the
lamina and the medulla. The third neuropile, the lobula, which is
located close to the cerebral lobe, is connected to the medulla by a long
(about 700 μm) nerve trunk (optic stalk) (*7*).* Thus, surgery between
the medulla and lobula can be easily made without damaging the lobula
region. We transected the optic stalk or removed the *optic lobe* bilaterally. Although the lobula region remained intact, the animals lost
their rhythmicity both in nymphal and adult stages (Fig. 2B), suggest-

---

*In this paper, the swollen part of optic lobe containing lamina and medulla is referred to
as *optic lobe*.

ing that in *G. bimaculatus* not the lobula but the medulla and/or lamina region play a crucial role for the manifestation of both the nymphal and adult rhythms (*23*).

Since these lesion experiments were always designed to find treatments which extinguished the circadian activity rhythm, the changes could have been caused by damage to crucial parts of the circadian pacemaker but not to the pacemaker itself, which could be situated somewhere outside the optic lobes. In other words, the lesion experiment does not provide conclusive proof for the hypothesis that the optic

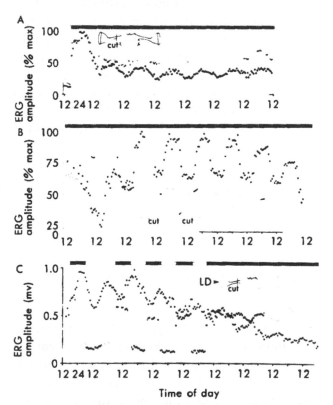

Fig. 3. A: circadian electroretinogram (ERG) amplitude rhythm in a cricket after unilateral severance of optic stalk (*22*). ○ operated (right) eye; ● intact (left) eye. B: circadian ERG rhythm under DD in a cricket after bilateral severance of optic stalks (*22*). ○ right eye; ● left eye. C: results of application of an entraining signal to the retina (○ right eye) after the optic stalk was sectioned (*24*). The contralateral eye (●) was intact and in DD. Note that the ERG rhythms showed an almost antiphase relationship in DD. ■■■ darkness; ☐ light.

lobe contains a self-sustained oscillator. The recent transplantation and regeneration experiments of Page (12–14) lend strong, though still indirect support to this hypothesis. Transplantation of optic lobes to a lobeless (therefore arrhythmic) cockroach led to a restoration of the rhythmicity with τ of donor. Furthermore, the circadian rhythm regained after optic tract severance showed values for τ and phase which were correlated with those of the preoperative rhythm.

Clearer indications that the *optic lobe* does in fact contain a complete circadian oscillator were provided by our electroretinogram (ERG) experiments in *G. bimaculatus* (22, 24). In intact animals, the ERG amplitude changed synchronously with LD, peaking in the dark fraction. This rhythm free-ran in the ensuing DD with a period somewhat shorter than 24 hr. This observation suggests that the rhythm is endogenous. Neither unilateral nor bilateral severance of the optic stalks affected the rhythm (Fig. 3A, B) (22). When the optic stalk was cut unilaterally, a 24-hr LD cycle provided to the eye of the operated side entrained the ipsilateral eye, but not the contralateral intact eye kept in DD. Consequently, the ERG rhythms of the two eyes were desynchronized (Fig. 3C) (24). These results indicate that the neurally isolated "*optic lobe-compound eye*" system contains the complete circadian system for the ERG rhythm (24). An important and still unresolved question is whether the ERG rhythm and the overt locomotor rhythm share the same pacemaker.

IV. CIRCADIAN RHYTHM IN AFFERENT NEURONAL ACTIVITY OF *OPTIC LOBE*

If the *optic lobe* is not merely a part of a pacemaker but contains the entire pacemaker controlling the overt rhythms, its afferent output should be rhythmic. This idea prompted an investigation of the afferent neuronal activities of the *optic lobe*. The afferent neuronal activity was recorded extracellularly from the distal cut end of the optic stalk using a suction electrode. The multiple unit activity (MUA) of *optic lobe* afferents exhibited two types of circadian rhythms. Type I showed a waveform like an unroofed trapezoid lacking its roof during the nighttime in LD cycles (Fig. 4A, B), and a sinusoidal waveform peaking in the subjective day under DD (Fig. 4B). Type II formed a characteristic

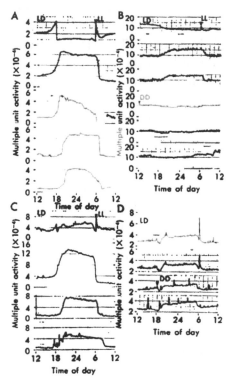

Fig. 4. Four examples of circadian rhythm of the afferent neuronal activity recorded from the distal cut end of the optic stalk. Each bin represents the total number of spike discharges per 5 min. Records continue from the top to the bottom. A, B: type I. C, D: type II. In B and C, the cerebral lobe was removed. In A, C, and D, animals were exposed to LD 12:12 (D: 18:00–6:00) for 1 (A, C) or 2 (D) days, then held in LL (A, C) or DD (D). In B, the animal was first exposed to LD 12:12 for 1 day and LL for 2 days, then held in DD. Under LL both types showed a characteristic trapezoid pattern in the subjective night (A–C). However, in DD, a phase-difference was seen. Type I peaked in the subjective day (B), while type II did in the subjective night (D). These results were obtained from more than 40 *optic lobes*. For further explanations see text.

trapezoid in the (subjective) night both in LD cycles and in DD (Fig. 4C, D). Under LL conditions, however, both types free-ran with a larger trapezoid in the subjective night (Fig. 4A–C). Removal of the subesophageal ganglion and/or cerebral lobe did not affect these rhythms (Fig. 4B, C). Even after optic nerve severance, the MUA showed either of the two rhythms just as in specimens with intact optic nerves kept in DD.

These two types of MUAs seem at least partly to be composed of different kinds of units. Both of them consist of many non-visual units firing autonomously, as well as of visual units. Type II MUA appears to contain more units firing spontaneously in the dark than type I. Although we could not analyze the responsiveness of each unit to light, according to Honegger's work (6), the dark firing units are thought to include three types: off-, dark firing on- and non-visual autonomously firing units.

The sensitivity of light on-units is evidently under control of the circadian pacemaker, as in the crayfish sustaining fibers (1). This is obvious from the fact that the characteristic trapezoidal wave has a larger amplitude under LL than under other lighting conditions (Fig. 4). The nocturnality of light on-units shows a parallel with the circadian ERG amplitude rhythm peaking in the night (22). Whether the off-units are controlled by the circadian pacemaker is still unclear. The on-units having spontaneous dark discharge, off-units and some kinds of non-visual units are all possible components to cause type II MUA rhythm in DD.

The results strongly suggest that the *optic lobe* generates a circadian rhythm and contains the pacemaker. However, since *optic lobes* were not isolated from humoral factors, we must discuss the possibility of a humoral control from circadian oscillator outside the optic lobe. Actually, some evidence has been obtained for oscillatory circadian structures located in the cerebral lobe of crickets (3, 16, Tomioka unpublished data). But no humoral factors from these oscillatory structures seem to participate in generating the *optic lobe*'s rhythmicity, since the circadian MUA rhythm persisted even after removal of the cerebral lobe. Although the possibility of humoral influence from some unknown oscillator elsewhere cannot be totally excluded, the simplest interpretation of the results along with optic lobe lesion and ERG experiments of many workers is that the *optic lobe* contains the circadian pacemaker which exerts its influence through afferent neurons on the central brain mass.

A question to be answered is what kind of units transmit the temporal information necessary for overt rhythm manifestation. Of course, the answer depends on future studies, but the results from the experiments on activity rhythms offer a cue for a solution. Both under LL and

DD, the locomotor rhythm of the cricket free-runs consistently, and the daily activity pattern is basically identical (21). Moreover, bilateral transection of the optic nerves did not disturb the activity rhythms (8, 20, 23). These facts suggest that the visual units are not essentially concerned with the overt rhythm generation; probably, they may be involved in the masking effect of light (21). It is therefore likely that some non-visual units mediate the pacemaker's output to the cerebral lobe.

In conclusion, we have obtained almost unequivocal evidence that, in the cricket (G. bimaculatus), the circadian pacemaker resides in the optic lobe and that its temporal information is carried to the driven system by certain optic lobe afferent neurons. However, much work remains to be done to completely understand the circadian system underlying the rest-activity rhythm. The structural elucidation of the circadian pacemaker itself, and the anatomical and physiological analysis of the output pathway from pacemaker to effector are two major problems for further studies.

SUMMARY

In insects, circadian pacemakers controlling the rest-activity rhythms have been roughly localized in some parts of the brain such as the optic lobe (cockroaches and crickets) and the cerebral lobe (silkmoths, mosquitoes, and houseflies). However, neither tissue has been demonstrated to contain the self-sustained oscillator.

We carried out a series of experiments in the cricket (G. bimaculatus) to localize the circadian pacemaker and to characterize the primary temporal information that is necessary for overt rhythm manifestation. Bilateral sectioning of optic stalks abolished the activity rhythm. The same intervention did not disturb the circadian ERG rhythm. These facts suggest that the optic lobe containing lamina and medulla contains the circadian pacemaker. Representing almost conclusive evidence for the optic lobe's self-oscillating ability, the circadian rhythmicity was detected in the afferent neuronal activity of the optic lobe that had been neurally isolated from the retina and other CNS structures. Thus, there is little doubt that the pacemaker in the optic lobe, whose actual structure should be elucidated in future studies, influences the rest-activity rhythm *via* a neural pathway.

Acknowledgments
I wish to thank Prof. Yoshihiko Chiba for many helpful discussions and critical reading of the manuscript. This work was supported by a grant from the Ministry of Education, Science and Culture of Japan (No. 59740366).

REFERENCES

1 Aréchiga, H. and Wiersma, C.A.G. (1969). *J. Neurobiol.* **1**, 71–85.
2 Chiba, Y. and Kasai, M. (1984). *Abstr. 10th Int. Congr. Biometeorol.*, p. 160.
3 Cymborowski, B. (1981). *J. Interdiscipl. Cycle Res.* **12**, 133–140.
4 Handler, A.M. and Konopka, R.J. (1979). *Nature* **274**, 708–710.
5 Helfrich, C., Cymborowski, B., and Engelmann, W. (1984). *Abstr. 10th Int. Congr. Biometeorol.*, p. 162.
6 Honegger, H.-W. (1978). *J. Comp. Physiol.* **125**, 259–266.
7 Honegger, H.-W. and Schürmann, F.W. (1975). *Cell Tissue Res.* **159**, 213–225.
8 Loher, W. (1972). *J. Comp. Physiol.* **79**, 173–190.
9 Nishiitsutsuji-Uwo, J. and Pittendrigh, C.S. (1968). *Z. Vergl. Physiol.* **58**, 1–13.
10 Nishiitsutsuji-Uwo, J. and Pittendrigh, C.S. (1968). *Z. Vergl. Physiol.* **58**, 14–46.
11 Page, T.L. (1978). *J. Comp. Physiol.* **124**, 225–236.
12 Page, T.L. (1982). *Science* **216**, 73–75.
13 Page, T.L. (1983). *J. Comp. Physiol.* **152**, 231–240.
14 Page, T.L. (1983). *J. Comp. Physiol.* **153**, 353–363.
15 Page, T.L., Caldarola, P.C., and Pittendrigh, C.S. (1977). *Proc. Natl. Acad. Sci. U.S.* **74**, 1277–1281.
16 Rence, B. and Loher, W. (1975). *Science* **190**, 385–387.
17 Roberts, S.K. (1965). *Science* **148**, 958–959.
18 Roberts, S.K. (1974). *J. Comp. Physiol.* **88**. 21–30.
19 Sokolove, P.G. (1975). *Brain Res.* **87**, 13–21.
20 Sokolove, P.G. and Loher, W. (1975). *J. Insect Physiol.* **21**, 785–799.
21 Tomioka, K. and Chiba, Y. (1982). *J. Comp. Physiol.* **147**, 299–304.
22 Tomioka, K. and Chiba, Y. (1982). *Naturwissenschaften* **69**, 395–396.
23 Tomioka, K. and Chiba, Y. (1984). *Zool. Sci.* **1**, 375–382.
24 Tomioka, K. and Chiba, Y. (1985). *J. Interdiscipl. Cycle Res.* **16**, 73–76.
25 Truman, J.W. (1974). *J. Comp. Physiol.* **95**, 281–296.

DELTA-SLEEP-INDUCING PEPTIDE

12

DSIP-LIKE MATERIAL IN RAT BRAIN, HUMAN CEREBROSPINAL FLUID, AND PLASMA AS DETERMINED BY ENZYME IMMUNOASSAY

NOBUMASA KATO,[*1] SHIGERU NAGAKI,[*1]
YASURO TAKAHASHI,[*2] IKURO NAMURA,[*3]
AND YOICHI SAITO[*4]

*Division of Diagnostics, National Center for Nervous, Mental and Muscular Disorders, Kodaira, 187,[*1] Department of Psychology, Tokyo Metropolitan Institute for Neurosciences, Fuchu 183,[*2] National Institute of Mental Health, Ichikawa, Chiba 272,[*3] and Department of Psychiatry, University of Tokyo, Tokyo 113,[*4] Japan*

Delta-sleep-inducing peptide (DSIP) was first isolated from the hemo-dialysate of torcular venous blood of rabbits in which the ventromedial laminar thalamus had been electrically stimulated (*11, 17*). Since then, DSIP has been proposed to be a natural sleep-promoting factor as evidenced by its enhancement of slow wave and/or REM sleep in several species (*5, 11, 13, 14*), though some reports appear to be skeptical about its sleep-promoting action (*3, 20*). We have reported that DSIP increased sleep on intraventricular infusion in the untreated dog compared to the saline-treated dog (*19*).

Kastin *et al.* (*7*) first reported a radioimmunoassay (RIA) of DSIP in the rat brain. The presence of the peptide has been demonstrated in dog plasma, cerebrospinal fluid (CSF) (*1*), human plasma (*6*), and milk (*4*). More recently, an immunohistochemical study (*2*) has shown the localization of DSIP pathways in the rat brain. We have reported an enzyme immunoassay (EIA) method of DSIP in dog plasma (*8*) and the regional distribution of immunoreactive DSIP (IR-DSIP) in the rat brain by EIA (*12*). This review comprises the presence of DSIP-like material in the brain and plasma of the rat, in the brain, CSF and plasma of man as well as some data on the 24-hr rhythm of free DSIP in human plasma as determined by EIA.

I. EIA OF DSIP

We have previously reported the development of EIA methods for thyroid-stimulating hormone (*9, 10*) and myelin basic protein (*18*) with the use of horseradish peroxidase as the enzyme label. The EIA method of DSIP has been developed by a similar technique (*8*). Briefly, specific antisera, which proved to be quite specific for the C (Glu)-terminus of the peptide, were raised in rabbits by immunizing with synthetic DSIP coupled with bovine serum albumin. The procedure of EIA was basically identical to that of competitive RIA except for the use of the enzyme-labeled DSIP instead of the radioactive tracer in RIAs. The bound and free fractions were separated by a double antibody-solid phase method using immunobeads coated with the purified second antibody. Enzyme activity bound to the bead was measured automatically

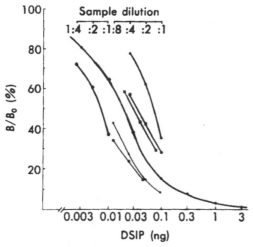

Fig. 1. Comparison of the EIA-DSIP standard curve with the dilution displacement curves of the samples from rat brain (△) and plasma (▲), and human CSF (■) and plasma (●). Pooled rat and human plasma were extracted, lyophilized, and reconstituted. Rat brain was extracted with 0.1 N acetic acid, whereas pooled CSF received no treatment prior to the concentration. B/B_0 is the ratio of the enzyme activity of the bound conjugates in the presence of a certain amount of DSIP (*B*) to that when no DSIP is present (B_0).

by fluorophotometry using an Auto FP-1 model photometer (Fuji-Rebio Inc.). Total incubation time was 18–24 hr at 4°C.

Figure 1 shows a typical standard curve for DSIP in the EIA over a range of 0.003–3 ng/tube. The assay sensitivity was 3 pg/tube at a 80% B/B_0 level and the DSIP concentration at 50% B/B_0 was 20 pg/tube. The nonspecific binding of the assay was 1.4%. The intra- and interassay coefficients of variation were 6.3 and 10.0%, respectively. Figure 1 also demonstrates that the assay of serially diluted samples taken from rat brain and plasma, and human CSF and plasma yielded displacement curves parallel to that of the DSIP standards. The results suggest that the present EIA reliably measures DSIP and that DSIP-like material is present in rat and human tissues. Assay time is relatively short and centrifugation is not required for the separation of the bound and free fractions. The enzyme-DSIP conjugate can be stored for over a year with little loss of activity. This EIA is therefore a simple and sensitive assay for DSIP without radiation hazards.

II. DSIP IN RAT BRAIN

The presence and molecular heterogeneity of IR-DSIP in the rat brain

TABLE I

Regional Distribution of IR-DSIP Concentrations in Rat Brain[a]

Region	IR-DSIP (ng/g tissue wet wt.)
Cerebellum	0.86 ± 0.04
Pons-medulla	0.71 ± 0.11
Striatum	0.81 ± 0.14
Nucleus septum	2.30 ± 0.48
Amygdala	2.18 ± 0.12
Hypothalamus	2.13 ± 0.09
Thalamus	1.88 ± 0.21
Midbrain	1.03 ± 0.23
Cerebral cortex	0.66 ± 0.13
Hippocampus	1.48 ± 0.23
Nucleus accumbens	3.21 ± 0.12
Piriform cortex	2.81 ± 0.13
Entorhinal cortex	2.53 ± 0.06

[a] Rats were killed by microwave irradiation and the brains were dissected on ice into 13 regions. Each region was weighed, homogenized by sonication in 0.1 N acetic acid, boiled and centrifuged twice. Aliquots of supernatant were lyophilized, reconstituted with the assay buffer and subjected to EIA. Values expressed as mean ± S.E.M. (12).

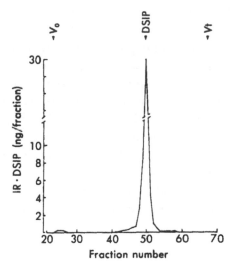

Fig. 2. Pooled brain extracts were reconstituted with 1 ml of 1 N acetic acid and applied onto a 1.5×75 cm Sephadex G-25 column eluted with 1 N acetic acid. Two-ml fractions were evaporated, reconstituted with the buffer and assayed by EIA. The column was calibrated with blue-dextran (V_0), synthetic DSIP, and methylene blue (V_t) (21).

were further investigated. Seven male Wistar rats were sacrificed and the brains were dissected and extracted with 0.1 N acetic acid. In the concentrated samples of brain extracts DSIP was measured by EIA. The regional quantitative distribution of IR-DSIP in the rat brain is presented in Table I. It varied 5-fold from 3.2 ng/g wet weight in nucleus accumbens to 0.7 ng/g in the cortex or pons-medulla though the contents were generally at a low level. IR-DSIP proved to be more concentrated in the limbic structures, which is consistent with the immunohistochemical findings (2).

The gel-filtration profile of the DSIP-like material in the rat brain has revealed that the major peak, representing 95% of the total activity, coeluted with authentic DSIP with only a minor peak eluting in the higher-molecular-weight fractions (Fig. 2). The predominance of the "free-form" of DSIP in rat brain has been noted in a previous RIA study (7).

Since total sleep deprivation has been used as a model for studying the brain metabolism and is known to alter serotonin metabolism (21), we have studied the effect of 24-hr sleep deprivation and subsequent rebound sleep on the IR-DSIP contents in the rat brain (12). The rats were deprived of sleep by handling for 24 hr starting at 10:00. Immediately after the 24-hr sleep deprivation, 5 rats were killed (sleep-deprived group), and 5 rats were allowed to sleep for 30 min (rebound-sleep group). Five rats were kept undisturbed and served as controls. The brain IR-DSIP showed neither in the sleep-deprived group nor in the 30-min rebound-sleep group a significant change. This negative result may cast some doubt on the involvement of endogenous DSIP (or "free" DSIP) in sleep regulation. However, some problems need to be further clarified, *i.e.*, the procedures to deprive of sleep inevitably cause stress and many peptides including DSIP are known to be influenced by stress (*16*). The procedures used in the study have been shown to increase REM sleep markedly, whereas they increased slow-wave sleep to a smaller extent (*21*).

III. DSIP IN HUMAN PLASMA

DSIP has been proposed to enhance delta and spindle electroencephalogram (EEG) patterns on intravenous infusion in rabbits (*11*) and the concentration of plasma DSIP-like material determined by RIA has been reported to be as high as 2–4 ng/ml in humans (*6*). In a preliminary study, we attempted to assay directly unextracted plasma taken from undisturbed and sleep-deprived dogs but failed to demonstrate the presence of endogenous DSIP. Although the EIA at the earlier stage

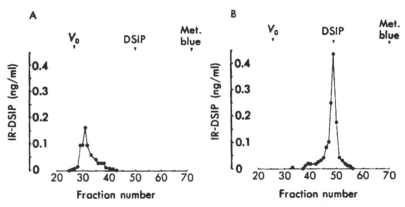

Fig. 3. The gel-filtration (Sephadex G-25) profiles of human plasma. Pooled plasma was treated with charcoal (A) or extracted with acetone and ethyl ether (B). The procedures are the same as in Fig. 2.

TABLE II
DSIP-like Material in Human Plasma

	at 11:00	14:00	17:00
Male (4)	25.2±1.2	28.3±1.0	30.5±2.1*
Female (4)	27.5±1.5	29.2±2.0	31.3±1.8
Total (8)	26.4±1.0	28.8±1.0	30.9±1.3**

Values; mean±S.E.M., expressed as pg/ml. Plasma extracted with acetone and ethyl ether. Comparison to 11:00 value: *$p < 0.05$, **$p < 0.01$.

was much less sensitive as compared to the present method, the reported level of 2–4 ng/ml should have been quite safely detected. Pooled plasma obtained from monkeys, dogs, and humans was treated with charcoal to obtain plasma from which any free-DSIP, but not bound-DSIP, had been removed. The dose-displacement curve generated by DSIP added in charcoal-treated plasma was virtually identical to the curve obtained with DSIP in assay buffer (8). These results suggest the bound-form DSIP is not detected by this method.

The column chromatography of human plasma treated with charcoal (Fig. 3A) was compared with that of plasma extracted with acetone and ethyl ether (Fig. 3B). The amount of bound-form DSIP was less than 60% of free-form DSIP by this EIA. This markedly contrasts with the observations by Kastin et al. (6) in which 89% of total DSIP-like material in plasma was bound to protein and 9% was present in free form.

These findings prompted us to assay extracted and concentrated samples from human plasma. Heparinized blood was taken from healthy volunteers and immediately mixed with EDTA and aprotinin. The harvested blood was centrifuged at 4°C and the plasma stored at −80°C until assayed. Prior to the assay, each plasma (1–2 ml) was extracted with 2 volumes of acetone and then with 4 volumes of ethyl ether. The extract was lyophilized, reconstituted with a minimum amount of buffer and subjected to EIA. The yield of synthetic DSIP added to plasma and extracted as unknown samples was 68±6%. The IR-DSIP values in these samples are supposed to reflect the amount of free-DSIP. The levels of IR-DSIP obtained in 8 subjects at 11:00, 14:00, 17:00 are shown in Table II. The samples taken at 17:00 contained significantly more free-DSIP immunoreactivity than those at 11:00. This finding was consistent with the previous results measuring preferentially bound-form DSIP by RIA (6). Plasma samples taken from 3 newborn babies with mild icterus were also assayed and they were proved to have more IR-DSIP (51±4 pg/ml) than adults. The assay of rat plasma withdrawn at 17:00 revealed that the level of IR-DSIP at 35±1.0 pg/ml ($n=6$) is comparable to human plasma.

IV. 24-HR RHYTHM OF FREE-DSIP IN HUMANS

The findings that the free-DSIP-like immunoreactivity increased between 11:00 and 17:00 (Table II) may suggest the presence of a circadian rhythm of IR-DSIP in humans. This possibility was further studied in 4 healthy male volunteers. An indwelling catheter was fixed in the cubital vein and heparinized blood was withdrawn at 30-min intervals during waking and at 20-min intervals during sleep throughout 24 hr. The EEG was simultaneously recorded during sleep. The

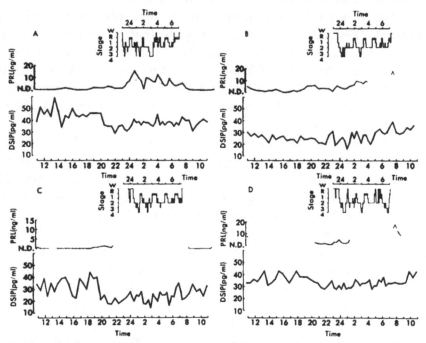

Fig. 4. Simultaneous 24-hr profiles of plasma prolactin (PRL) (middle panel) and of IR-DSIP in extracted plasma (bottom panel) obtained in 4 healthy male subjects (A, B, C, and D). Sleep stages are represented in the top panel.

levels of free-DSIP were determined by the assay of plasma extract with acetone and ether. Plasma prolactin (PRL) concentrations were also determined by RIA and compared with IR-DSIP levels. The kit for RIA-human PRL was kindly supplied by NIADDK, USA.

Figure 4 illustrates the 24-hr profiles of plasma PRL and of IR-DSIP in extracted plasma with sleep histograms. Plasma PRL exhibited higher values during sleep than during daytime, as is well-known (*15*). IR-DSIP levels, which mostly represent free-DSIP, showed considerable day-to-day and individual variations. It is difficult to reach a definite conclusion from these results, although IR-DSIP levels generally tend to decrease in nighttime. There are indications of a pulsatile pattern of DSIP-secretion. It is still unclear whether the secretion of endogenous DSIP is sleep-related and dependent on a specific stage of the sleep histogram.

V. DSIP IN HUMAN CEREBROSPINAL FLUID

The dog CSF has been reported to contain DSIP-like material at a level of 0.56 ± 0.03 ng/ml, of which 44% eluted immediately after the void

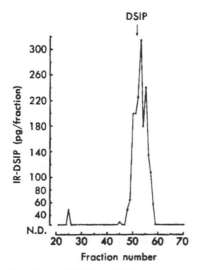

Fig. 5. A Sephadex G-25 chromatography of human untreated CSF. Pooled CSF was centrifuged, the supernatant evaporated into dryness under vacuum and otherwise treated as in Fig. 2.

volume (bound fraction), 51% coeluted with DSIP (free-DSIP) and a small percent coeluted with desTrp^1-DSIP (free des-Trp^1-DSIP) (*1*). We have attempted to detect endogenous DSIP in the human CSF. CSF samples taken from patients with various disorders were extracted with acetone and ether and assayed by EIA ("extracted"). The aliquots of several samples were centrifuged and the supernatants lyophilized and subjected to the assay ("untreated"). The IR-DSIP determinations in CSF showed fairly constant values irrespective of age and disorders. The fact that some of the CSF samples had been left at room temperature for a while after the puncture seemed not to affect the values. The IR-DSIP in extracted CSF from children was 44 ± 0.8 pg/ml ($n=8$) and in CSF from adult patients 44 ± 1.5 pg/ml ($n=11$). These values correspond to 60–80% of the values in untreated CSF, and seem to be reasonable considering the yield of the extraction procedures. The column (Sephadex G-25) chromatography of pooled and concentrated CSF exhibited more than 90% of total IR-DSIP activity coeluting with authentic DSIP (Fig. 5), a finding that again is inconsistent with previous results in the dog CSF (*1*).

VI. DSIP IN HUMAN BRAIN

There are no indications as to the presence of DSIP in human brain which is known to contain many neuropeptides. Small samples of brain tissue from 2 patients, generously provided by Dr. H. Kaiya, Department of Psychiatry, Gifu University, served to investigate this. The extraction and assay procedures followed the method for rat brain.

Three adjacent tissues in the frontal pole (Broadman's Area 10) contained IR-DSIP at concentrations of 1.3, 1.6, and 1.5 ng/g tissue wet weight, and 1.9 ng/g in the inferior frontal gyrus (Area 44) of a 72-year old female who had died of breast cancer. The temporal pole (Area 38) of a 65-year old male patient, who had died of lung cancer, showed a higher IR-DSIP estimate of 9.5 ng/g, but this value decreased to 2.1 ng/g after acetone-ether extraction. The IR-DSIP levels seem to be comparable or somewhat higher than those in rat brain (Table I). An aliquot of the latter sample was applied onto a Sephadex G-50 column eluted with 1 N acetic acid. As shown in Fig. 6, the gel-filtration profile presented a striking contrast to our results in human plasma, CSF and

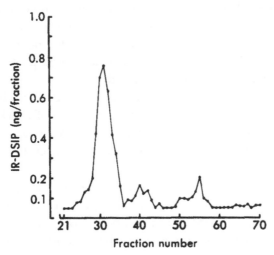

Fig. 6. A gel-filtration profile of human brain extract (Area 38) from a 65-year old male patient who died of lung cancer. Nine ml (0.9 g tissue) of the brain extract was lyophilized and applied onto a Sephadex G-50 column as described in Fig. 2. Note that Sephadex G-50 was used with the same bed volume as in Figs. 2, 3, and 5.

rat brain mentioned before. Almost all of total IR-DSIP activity was eluted immediately after the void volume and no peak was found in the fractions coeluting with DSIP. This peak may suggest the bound form of DSIP, but its nature is not known.

Although this is a preliminary finding, DSIP-like material with higher molecular weight than authentic DSIP seemed to be present in human brain. This apparent discrepancy between human brain and other tissues deserves further investigation.

VII. CONCLUSION

Our estimates of DSIP-like material in rat and human tissues by EIA are much lower (100 times lower in plasma and 10 times in CSF) than those by RIA in previous reports (1, 6, 7). The findings of column chromatography suggest that the material detected by this EIA is mainly the free-form DSIP and/or unknown substances with similar molecular weight. This striking difference between our observations and those of Kastin et al. (1, 6, 7) might be best explained in terms of the difference in antibodies used (8. 12). Our antibody can only recognize free-DSIP, whereas the antibody used in RIA studies may detect both bound and free peptides. The antibody used in this study could be diluted up to 1:80,000 and was highly specific for the C-terminus of the peptide. This indicates the possibility that DSIP is bound to a carrier protein in the C-terminus, the recognition site of the antibody.

Preliminary results obtained for the human brain appear to be contradictory to the above explanations. The possible presence of DSIP-like material with high molecular weight in human brain tissue cannot be readily explained, but may reflect some species differences to be further elucidated.

Assuming some analogy with the thyroid hormone, one can suppose that the free-DSIP is an active form and the bound peptide a reservoir. In support of this assumption, free-DSIP seems to be a more dominant form in the central nervous system as shown for rat brain, and dog and human CSF, as compared to plasma.

While DSIP has been suggested to be a natural sleep-promoting factor mainly on the basis of neurophysiological techniques, studies based on immunoassays have only recently been undertaken. Our studies concerning sleep deprivation failed to show a significant change in brain IR-DSIP content, and also the profiles of 24-hr IR-DSIP levels in human plasma are still ambiguous with respect to the sleep-wakefulness cycle. However, the presence of endogenous DSIP was established in rat and human tissues in this study. The present EIA method, which preferentially detects free-DSIP, may provide a good tool for exploring the possible roles of endogenous DSIP.

SUMMARY

A method for the EIA of DSIP has been developed. The method was applied to investigate the DSIP-like material in rat brain and plasma, and human brain, CSF, and plasma as well as the 24-hr rhythm of DSIP in human plasma. The presence of DSIP-like material has been confirmed in all body-fluids and tissues tested. Most of the DSIP-immunoreactivity by this enzyme immunoassay proved to be free-form DSIP except for the human brain in which DSIP-like material with higher molecular weight was suggested to be dominant.

The 24-hr sleep deprivation and subsequent rebound sleep had no significant effect on rat brain contents of DSIP-like material. The profiles of 24-hr free-DSIP levels in human plasma were compared with plasma prolactin concentrations and sleep histograms. Plasma DSIP levels failed to show distinct circadian rhythms with respect to the sleep-wakefulness cycle, though they tended to be lower in the nighttime.

Acknowledgments
We wish to thank The Pituitary Agency, NIADDK (USA) for the supply of a hPRL-Kit and Dr. H. Kaiya, Department of Psychiatry, Gifu University for the generous gift of brain samples. We are grateful to Fuji-Rebio Inc. (Tokyo) for supplying the materials for EIA. The technical help of Ms. N. Watanabe is appreciated. This work was supported in part by grant (No. 58570498) from the Ministry of Education, Science and Culture, Japan.

REFERENCES

1 Banks, W.A., Kastin, A.J., and Coy, D.H. (1982). *Pharmacol. Biochem. Behav.* **17**, 1009–1014.
2 Constantinidis, J., Bouras, C., Guntern R., Taban, C.H., and Tissot, R. (1983). *Neuropsychobiology* **10**, 94–100.
3 Drucker-Colin, R. (1981). In *Psychopharmacology of Sleep*, ed. Wheatley, D., pp. 53–72. New York: Raven Press.
4 Graf, M.V., Hunter, C.A., and Kastin, A.J. (1984). *J. Clin. Endocrinol. Metab.* **59**, 127–132.
5 Kafi, S., Monnier, M., and Gallard, J.M. (1979). *Neurosci. Lett.* **13**, 169–172.
6 Kastin, A.J., Castellanos, P.F., Banks, W.A., and Coy, D.H. (1981). *Pharmacol. Biochem. Behav.* **15**, 969–974.
7 Kastin, A.J., Nissen, C., Schally, A.V., and Coy, D.H. (1978). *Brain Res. Bull.* **3**, 691–695.
8 Kato, N., Honda, Y., Ebihara, S., Naruse, H., and Takahashi, Y. (1984). *Neuroendocrinology* **39**, 39–44.
9 Kato, N., Ishii, S., Naruse, H., Irie, M., Arakawa, H., and Tsuji, A. (1980). *Anal. Lett.* **13**, 1555–1565.
10 Kato, N., Naruse, H., Irie, M., and Tsuji, A. (1979). *Anal. Biochem.* **96**, 419–425.
11 Monnier, M., Dudler, L., Gächter, R., and Schoenenberger, G.A. (1977). *Neurosci. Lett.* **6**, 9–13.
12 Nagaki, S. and Kato, N. (1984). *Neurosci. Lett.* **51**, 253–257.
13 Nagasaki, H., Kitahama, K., Valatx, J.-L., and Jouvet, M. (1980). *Brain Res.* **192**, 276–280.
14 Polc, P., Schneeberger, J., and Haefely, W. (1978). *Neurosci. Lett.* **9**, 33–36.
15 Sassin, J.F., Frantz, A.G., Kapen, S., and Weitzman, E.D. (1973). *J. Clin. Endocrinol. Metab.* **37**, 436–440.

16 Schneider-Helmert, D. and Schoenenberger, G.A. (1983). *Neuropsychobiology* 9, 197–206.
17 Schoenenberger, G.A. and Monnier, M. (1977). *Proc. Natl. Acad. Sci. U.S.* 74, 1282–1286.
18 Shinomiya, Y., Kato, N., Imazawa, M., and Miyamoto, K. (1982). *J. Neurochem.* 39, 1291–1296.
19 Takahashi, Y., Kato, N., Nakamura, Y., Ebihara, S., Tsuji, A., and Takahashi, K. (1980). In *Integrative Control Functions of the Brain*, ed. Ito, M., Tsukahara, N., Kubota, K., and Yagi, K., vol. 2, pp. 339–341. Amsterdam: Elsevier.
20 Tobler, I. and Borbély, A.A. (1980). *Waking Sleep.* 4, 139–153.
21 Toru, M., Mitsushio, H., Mataga, N., Takashima, M., and Arito, H. (1984). *Pharmacol. Biochem. Behav.* 20, 757–761.

13

THE NATURAL OCCURRENCE AND SOME EXTRA-SLEEP EFFECTS OF DSIP

MARKUS V. GRAF,* ABBA J. KASTIN,
AND ALAN J. FISCHMAN

VA Medical Center and Tulane University, School of Medicine, New Orleans, LA 70146, U.S.A.

Delta-sleep-inducing peptide (DSIP) was isolated and characterized by Schoenenberger and Monnier in 1976/1977 (*29*, *37*), presumably in its natural form. The effects of injected DSIP, however, were not as robust as many researchers expected (for a review see ref. *10*). As a consequence, the possibility arose that this peptide might not exist naturally, thereby explaining some failures to find effects in the early studies. It was possible that the structure of DSIP might have represented only a part of a larger polypeptide degraded during the initial procedure of isolation but still retaining some capacity to induce delta-wave sleep in rabbits. This capacity was used as the test-system in the isolation at DSIP from hemodialysates of rabbits electrically stimulated in the "sleep center" of the thalamus (*37*). It was also possible that the apparent release of DSIP might have been produced by direct electric cleavage of an ionic bond of the peptide with formation of an unknown compound essential for the full activity of DSIP. A clear demonstration

---

*Present address: Res. Dept., ZLF 403, Kantonsspital, CH-4031 Basel, Switzerland.

of the natural occurrence of a peptide identical in structure to the reported DSIP, therefore, would be of value.

I. DIFFERENT FORMS OF DSIP-LI

In 1978, we developed a radioimmunoassay (RIA) for DSIP using antibody No. 604 that was generated in rabbits immunized with a conjugate of DSIP and α-globulin (21). DSIP-like immunoreactivity (DSIP-LI) was measurable in rat brain after acid extraction of the tissue. Chromatography of the extract on a column of Sephadex G-25 and subsequent RIA of the fractions revealed a peak of DSIP-LI at the position where synthetic DSIP eluted. This was the first indication, other than the initial isolation, that DSIP occurred naturally. However, questions were again raised a few years later when we found that most of the DSIP-LI in plasma eluted near the void volume of the columns with only a minor peak at the elution position of DSIP (20). This pattern could have been caused by a large nonspecific interfering substance present in plasma. Acidification, however, of this large molecular size form (LF) of DSIP-LI again produced two peaks in a subsequent chromatography, one of the peaks eluting at the position of DSIP (20). This strongly suggested that a substance structurally related to DSIP was bound to or present in the material eluting with the void volume. Thus, two different forms of DSIP seemed to occur in plasma: "free" and "bound". At the same time, we showed that addition of the synthetic DSIP nonapeptide primarily increased the immunoreactivity in the first (LF) peak.

More questions were raised when Ekman et al. reported that they could not find any DSIP-LI at the position of DSIP after chromatography of plasma on Bio-Gel P-10 (6). The same absence was shown in urine where these authors only detected DSIP-LI at different elution positions, although in cerebrospinal fluid (CSF) one of the peaks apparently corresponded to DSIP. It is difficult to estimate the molecular weight or size of their first peaks because they did not indicate the void volume of the column, but the molecular weight of the DSIP-LI they found in plasma and urine (and partly in CSF) was greater than 850, the molecular weight of DSIP.

Additional evidence for multiple forms of DSIP in vivo was provided by Rozhanets *et al.* (*31*). These authors found DSIP-LI in brain that showed a similar chromatographic elution profile as DSIP, but different peaks of DSIP-LI appeared when extracts of kidney and liver were chromatogramed.

Other studies also suggested different forms of DSIP. It was observed (*10*), for instance, that DSIP labeled with tritium at Trp^1 disappeared from plasma within 5 min after intravenous administration, as evaluated by thin-layer chromatography (TLC). Although rapid degradation of DSIP in brain slices or homogenates with a half-life of about 15 min has been reported (*16, 26*), the disappearance in plasma seemed to occur at even a faster rate.

The possibility, however, was not excluded that ^{3}H-DSIP may have been bound or adsorbed to other molecules and, therefore, not detected at the location of synthetic DSIP by TLC. Fast disappearance-rates of DSIP in plasma of dog, monkey, and rat with a half-life of a few minutes were also reported recently by Kato *et al.* (*23*). Since these authors indicated that their assay system did not measure bound DSIP, the fast disappearance of DSIP in plasma could not only reflect degradation of the peptide but also a masking of the measurable compound *in vivo*.

Similar fast rates of disappearance were found in human milk, whereas the half-life of measurable DSIP-P (the phosphorylated analog) in breast milk was around 11 hr and no degradation at all was found by RIA for several hours in cow milk (*41*; G.A. Schoenenberger, personal communication). It is not known whether this represents a difference in enzymatic activity or a change in the masking ability of the different sources of milk.

II. DSIP IN LF-DSIP-LI

In a study designed to measure DSIP-LI in peripheral organs of the rat (*11*), we found that extraction with water yielded much higher values of DSIP-LI than extraction with acid, as was used in earlier studies with brain (*21*). Only a minor amount of this immunoreactivity (IR) appeared as the small molecular weight form (SF)-DSIP-LI after gel filtration on columns of Sephadex G-15 or G-25, whereas the majority eluted as the LF. Direct evidence for the existence of SF-DSIP-LI within the LF-DSIP-LI was obtained by incubation of the LF with trypsin. Such enzymatic treatment resulted in a "shift" of the main peak on Sephadex G-25 so that no DSIP-LI was found eluting with the void volume, but all of the IR eluted in the region of SF-DSIP-LI with a considerable amount in the fractions where synthetic DSIP appeared (11). Similar results were observed with spleen, liver, and jejunum.

Another indication for the existence of DSIP in different forms was seen when charcoal incubated with homogenates of the different organs was found to remove only negligible amounts of DSIP-LI at concentrations of 1–2 mg wet weight tissue per ml of homogenate (11). Only further dilutions of the homogenates and subsequent incubation with charcoal progressively removed higher percentages of the measured DSIP-LI. These results pointed to an equilibrium process in the homogenates between DSIP-LI and another compound; the equilibrium apparently shifted to a higher percentage of SF-DSIP-LI at lower concentrations. The nature of the LF-DSIP-LI that appeared to be non-adsorbable by charcoal is still unknown, but we believe that at least part of it contains the immunoreactive DSIP sequence.

In a different study involving the measurement and characterization of DSIP-LI in human milk, additional evidence for multiple forms of DSIP was obtained (9). Chromatography of breast milk on columns of Sephadex G-15 and G-10 revealed a large peak of LF-DSIP-LI in addition to SF-DSIP-LI. After LF-DSIP-LI was digested with trypsin and chromatogramed on a column of Sephadex G-10, all the IR appeared at the position of synthetic DSIP. High performance liquid chromatography (HPLC) of this peak produced a main peak that co-eluted with synthetic DSIP and a second peak that co-eluted with P-DSIP. Thus, for the first time, it was shown with the more sophisticated technique of HPLC that DSIP existed as the nonapeptide in a body fluid and to a large extent, was present in LF-DSIP-LI. In addition, this was also the first demonstration of the natural occurrence of DSIP-P. The existence of both DSIP and DSIP-P in milk was later confirmed by the group in Basel (41).

III. "FREE" DSIP

In the same series of experiments (9), the SF-DSIP-LI that was found after gel filtration of human milk was directly subjected to HPLC. With this method, we detected several peaks of DSIP-LI, two of them eluting at the position of DSIP and DSIP-P as determined by RIA. This indicated that apparently free DSIP and DSIP-P, in addition to the LF-DSIP-LI, were present in human milk.

Further studies concerning the natural occurrence of DSIP in plasma, CSF, and urine were aimed at the unequivocal demonstration of freely occurring DSIP in these fluids. By means of gel filtration on Sephadex G-100, HPLC, and analysis of the fractions by RIA, we demonstrated the presence of a compound in plasma of human, rabbit, rat, and dog that eluted with both separation techniques at the position of DSIP and also crossreacted with antiserum No. 607 in the RIA with DSIP (*14*). A similar finding was obtained with human CSF. These results indicate the existence of free DSIP in these body fluids.

IV. DSIP, A FOLDED CONFORMATION?

Antibody (Ab) No. 607 used in these experiments requires virtually the entire sequence of the nonapeptide for recognition. Des-Trp-DSIP and DSIP amidated at the C-terminal Glu cross-react little, if any, with this antibody. In contrast, DSIP-P was fully cross-reactive with DSIP Ab 607 as was D-Ala^4-DSIP, but D-Ala^3-DSIP was not (*14*). Based on spectroscopic analyses, it was proposed that DSIP in solution can exist in a folded conformation (*1, 18, 27*). Two of our observations support this proposition: i) Ab 607 does not recognize peptides related to DSIP with changes in the amino acids at either end but can recognize peptides with changes in the middle part of the sequence. This is consistent with a folded conformation of DSIP in which both ends of the molecule contribute to a single recognition site for the antibody; ii) there is a big difference between the crossreactivities of Ab 607 with D-Ala^3-DSIP and with D-Ala^4-DSIP. When D-Ala is placed in position 3, it is known from studies with proteins (*4*) that a II′ β-turn is favored, whereas the same residue placed in position 4 favors a II β-turn. Thus, if Ab 607 requires a type II β-turn conformation at this part of the sequence, its particular pattern of crossreactivity is explained. At the same time, this conformation brings the C- and N-terminal amino acids into close vicinity with each other, forming an apparently uniform epitope. Thus, there is experimental evidence in addition to that provided by spectroscopy that free DSIP in solution exists in a folded conformation.

This spatial arrangement conceivably would help explain the puzzling phenomenon of different laboratories finding different amounts of DSIP-LI in plasma. Values of DSIP-LI in human plasma have been reported to be almost 40 pg/ml (*30*), about 300 pg/ml (*14, 31*), and up to several ng/ml (*20, 41*). These considerable discrepancies may be explained in part by the particular determinants of the different antibody.

V. MEASURABLE DSIP-LI

Our own studies performed with the same basic methodology but with different antisera also yielded different results. It is, therefore, reasonable to conclude that different antibodies to DSIP recognize different amounts of the naturally occurring peptide according to the varying amounts of the different forms of DSIP. We found about 10 times less DSIP-LI in plasma with Ab 607 than with Ab 604 (*14*). Gel chromatography of plasma analyzed by RIA with Ab 604 showed a large peak of LF-DSIP-LI and only a minor peak of SF-DSIP-LI, whereas the same procedure yielded the opposite pattern with Ab 607. This indicates that Ab 607 recognizes mostly the free DSIP, whereas Ab 604 additionally detects a large amount of LF-DSIP-LI. Since it appears that Ab 607 probably reacts with the N- and C-termini of the folded DSIP structure and Ab 604 probably reacts with the opposite central side, it is possible that DSIP in the large form (LF-DSIP-LI) is attached by the N–C-terminal site, essentially masking the site to which Ab 607 binds.

It is not known if the equilibrium system of the large and small forms of DSIP is of importance for the biological effects of the peptide. Some peculiarities of DSIP seem to point to such an influence, for instance the inverted U-shaped dose-response curve (*22, 38*), the optimal time of infusion (*35*), the fast disappearance of exogenous DSIP in plasma (*23*), and the "protection against degradation" of endogenous DSIP-LI (G.A. Schoenenberger, personal communication). All these attributes of DSIP may depend on the ratio of "bound" to "free" peptide in blood and similar considerations may also apply to this peptide in brain as well as peripheral organs.

VI.   CIRCADIAN RHYTHM

Circadian variations in the level of DSIP-LI in plasma may be restricted to certain forms of DSIP. With Ab 607, we found a significant rhythm of DSIP-LI over 24 hr in rat plasma with a trough between 8:00 and 11:00 and a peak at 17:00, shortly before the onset of the dark-period at 18:00 (7). This rhythm correlated with the diurnal variation of corticosterone measured in the same rats, but the regulating factors seemed to be different since constant conditions of illumination influenced both parameters differently. The peak of DSIP-LI in rat plasma at 17:00 was rather surprising because peak-values at this time of day have also been reported in humans (20, 24). Since the organization of the circadian rhythms in humans and rats is generally reversed, the question arises whether DSIP fulfills a different function in both species. Sleep-induction by DSIP has been observed in both these species (17, 19, 34, 40), but the time of day necessary for injection of DSIP in order to obtain optimal effects seems to be different.

VII.   INTERACTION WITH AMPHETAMINE

We have discussed previously the possibility of DSIP being more than a sleep-peptide (22). It could be expected that a sleep-inducing factor would possess some additional actions that might be easier to evaluate than sleep. Accordingly, we investigated several such actions of DSIP in mice and rats in more detail.

We chose amphetamine-induced hyperthermia as a model to evaluate the effects of DSIP in mice (42). DSIP reduced the temperature increase induced by amphetamine with a maximal effect seen between 100 and 200 nmol/kg i.p. (13). The effects of DSIP as well as D-Ala^4-DSIP were similar, but DSIP-P apparently did not affect amphetamine-induced hyperthermia. The active dose ranges are comparable to those obtained by Scherschlicht et al. (33) in a different system. Surprisingly, a second effective dose range as low as 0.1 nmol/kg was found for DSIP. Such a finding is unusual and indicates that the dose-response relationship of DSIP may be more complex than with many other biologically active compounds.

Similar complex interactions of dose and effect were observed in mice when the influence of DSIP on amphetamine-induced locomotor activity was studied (*15*). A lower dose of DSIP (30 nmol/kg) reduced or increased locomotor activity depending on the concentration of amphetamine (10 or 15 mg/kg). A higher dose of DSIP (120 nmol/kg) decreased activity in both cases; however, after a latency of about 150 min, DSIP increased activity when combined with the higher dose of amphetamine. It is not known if this increase of activity was due to a presumably decreased concentration of DSIP at 150 min that produced effects comparable to those seen with 30 nmol/kg DSIP at earlier times.

VIII. PATHWAYS INVOLVED

No clear indications have been obtained so far as to the mechanism of action of DSIP. Several ideas have been put forward (*10*), but conclusive support is lacking. It is reasonable to assume an involvement of serotonin in the effects of DSIP as proposed by Yehuda and Mostofski (*43*), since this neurotransmitter has always been considered a major factor in sleep-events (*28*). An influence of DSIP on the content of serotonin in brain was found in an earlier study (*8*) where an effect on norepinephrine was also observed. Recently, we observed an interaction of DSIP with adrenergic receptors in an experiment in which DSIP inhibited the nocturnal increase of N-acetyltransferase (NAT) activity in rat pineal (*12*). This inhibition occurred after injection of DSIP in the evening with an optimal dose of 30 nmol/kg i.v. In this system, DSIP-P and D-Ala^4-DSIP were at least as active as DSIP. Since adrenergic regulation of NAT-activity is well established (*25*), this could indicate an involvement of the adrenergic system in the actions of DSIP. It is possible that the interactions of DSIP with amphetamine may also be related to this system since so many of the known effects of amphetamine appear to be mediated by release of catecholamines (*2, 5*).

IX. STRESS

Stress has been found in different studies to be influenced by DSIP. DSIP has been reported to normalize stressed rabbits in their sleep behavior (*32*), to make rats more resistant to electric shock (*39*), and to enhance tolerance against stress in humans (*36*). We found support for these findings in the influence of DSIP on corticotropin-releasing factor (CRF)-stimulated release of corticosterone in chlorpromazine-morphine-pentobarbital treated rats (submitted). The release of corticosterone was reduced by 5–30 μg/kg DSIP i.v. but not by doses of 1 or 50–100 μg/kg. This, again, reflects an inverted U-shaped dose-response curve. The same doses of DSIP did not affect the release of corticosterone induced by adrenocorticotrophic hormone (ACTH); this suggests that the interaction of DSIP with CRF probably occurs at the level of the pituitary. It is, however, not known if the effects on sleep are related to the apparent stress-reducing properties of DSIP.

SUMMARY

Since the isolation of DSIP, different methods have been used in attempts to demonstrate the natural existence of this peptide. By means of gel chromatography, some evidence for the existence of DSIP was obtained. By HPLC, we recently demonstrated the occurrence of "free" DSIP in plasma of different species as well as in CSF, urine, and milk of humans. By tryptic degradation of the "large molecules" and subsequent analysis by HPLC and RIA, we also showed that DSIP-like structures could be obtained from the large compounds. Some evidence exists for an interaction of DSIP with the serotonergic as well as adrenergic system, but the question whether DSIP is more than a sleep-peptide is still open.

REFERENCES

1 Akhrem, A.A., Galaktionov, S.G., Golubovich, R.V., and Kirnarskii, L.I. (1982). *Biophysics* 27, 334–335.

2 Altar, A., Terry, R.L., and Lytle, L.D. (1984). *Gen. Pharmacol.* 15, 13–18.

3 Axelrod, J. (1974). *Science* 184, 1341–1348.

4 Chou, P.Y. and Fasman, G.D. (1977). *J. Mol. Biol.* **115**, 135–175.

5 Costa, E. and Garattini, S. (1970). *Amphetamine and Related Compounds.* New York: Raven Press.

6 Ekman, R., Larson, I., Malmquist, M., and Thorell, J.I. (1983). *Regul. Pept.* **6**, 371–378.

7 Fischman, A.J., Kastin, A.J., and Graf, M.V. (1984). *Life Sci.* **35**, 2079–2084.

8 Graf, M., Baumann, J.B., Girard, J., Tobler, H.J., and Schoenenberger, G.A. (1982). *Pharmacol. Biochem. Behav.* **17**, 511–517.

9 Graf, M.V.. Hunter, C.A., and Kastin, A.J. (1984). *J. Clin. Endocrinol. Metab.* **59**, 127–132.

10 Graf, M.V. and Kastin, A.J. (1984). *Neurosci. Biobehav. Rev.* **8**, 83–93.

11 Graf, M.V. and Kastin, A.J. (1984). *Proc. Soc. Exp. Biol. Med.* **177**, 197–204.

12 Graf, M.V., Kastin, A.J., and Schoenenberger, G.A. (1984). *J. Neurochem.* **44**, 629–632.

13 Graf, M.V., Kastin, A.J., Coy, D.H., and Zadina, J.E. (1984). *Physiol. Behav.* **33**, 291–295.

14 Graf, M.V., Kastin, A.J., and Fischman, A.J. (1984). *Pharmacol. Biochem. Behav.* **21**, 761–766.

15 Graf, M.V., Zadina, J.E., and Schoenenberger, G.A. (1982). *Peptides* **3**, 729–731.

16 Huang, J.T. and Lajtha, A. (1978). *Res. Commun. Chem. Pathol. Pharmacol.* **19**, 191–199.

17 Inoué, S., Honda, K., Nishida, S., and Komoda, Y. (1983). IV. Int. Congr. APSS, Bologna, Abstracts, p. 212.

18 Ivanov, V.T., Mikhaleva, I.I., Sargsyan, A.S., Balashova, T.A., Efremov, E.S., Deshko, T.N., and Nabiev, I.R. (1981). In *Peptides 1980*, ed. Brunfeldt, K., pp. 501–507. Copenhagen: Scriptor.

19 Kafi, S., Monnier, M., and Gaillard, J.M. (1979). *Neurosci. Lett.* **13**, 169–172.

20 Kastin, A.J., Castellanos, P.F., Banks, W.A., and Coy, D.H. (1981). *Pharmacol. Biochem. Behav.* **15**, 969–974.

21 Kastin, A.J., Nissen, C., Schally, A.V., and Coy, D.H. (1978). *Brain Res. Bull.* **3**, 691–695.

22 Kastin, A.J., Olson, R.D., Schally, A.V., and Coy, D.H. (1980). *Trends Neurosci.* **3**, 163–165.

23 Kato, N., Honda, Y., Ebihara, S., Naruse, H., and Takahashi, Y. (1984). *Neuroendocrinology* **39**, 39–44.

24 Kato, N., Nagaki, S., Takahashi, Y., Namura, I., and Saito, Y. (1984). 8th Taniguchi Symp., Kyoto-Katata, Abstracts, pp. 28–29.

25 Klein, D.C., Sugden, D., and Weller, J.L. (1983). *Proc. Natl. Acad. Sci. U.S.* **80**, 599–603.

26 Marks, N., Stern, F., Kastin, A.J., and Coy, D.H. (1977). *Brain Res. Bull.* **2**, 491–493.

27 Mikhaleva, I., Sargsyan, A., Balashova, T., and Ivanov, V. (1982). In *Chemistry of Peptides and Proteins*, ed. Voelter, W., Wuensch, E., Ovchinnikov, J., and Ivanov, V., vol. 1, pp. 289–297. Berlin: de Gruyter.

28 Monnier, M. and Gaillard, J.M. (1980). *Experientia* **36**, 21–24.

29 Monnier, M. and Schoenenberger, G.A. (1977). In *Sleep 1976*, ed. Koella, W.P. and Levin, P., pp. 257–263. Basel: S. Karger AG.

30 Nagaki, S., Kato, N., Watanabe, N., Naruse, H., Namura, I., Saito, Y., and Takahashi, Y. (1983). IV. Int. Congr. APSS, Bologna, Abstracts, p. 210.

31 Rozhanets, V.V., Yukhananov, R.Y., Chizhevskaya, M.A., and Navolotskaya, E.V. (1983). *Neurokhimiya* **2**, 353–363. (in Russian)

32 Scherschlicht, R., Marias, J., Schneeberger, J., and Steiner, M. (1981). In *Sleep 1980*, ed. Koella, W.P. and Levin, P., pp. 147–155. Basel: S. Karger AG.

33 Scherschlicht, R., Aeppli, L., Polc, P., and Haefely, W. (1984). *Europ. Neur.* **23**, 346–352.

34 Schneider-Helmert, D., Gnirss, F., Monnier, M., Schenker, J., and Schoenenberger, G.A. (1981). *Int. J. Clin. Pharmacol. Ther. Toxicol.* **19**, 341–345.

35 Schneider-Helmert, D., Gnirss, F., and Schoenenberger, G.A. (1981). In *Sleep 1980*, ed. Koella, W.P. and Levin, P., p. 132. Basel: S. Karger AG.

36 Schneider-Helmert, D. and Schoenenberger, G.A. (1983). *Neuropsychobiology* **9**, 197–206.

37 Schoenenberger, G.A., Maier, P.F., Tobler, H.J., and Monnier, M. (1977). *Pflüegers Arch.* **369**, 99–109.

38 Schoenenberger, G.A. and Monnier, M. (1979). In *IUPAC Medical Chemistry Proceedings*, pp. 101–116. Oxford: Cotswold Press.

39 Sudakov, K.V., Ivanov, V.T., Koplik, E.V., Vedjaev, D.F., Michaleva, I.I., and Sargsyan, S. (1983). *Pavlov. J. Biol. Sci.* **18**, 1–5.

40 Ursin, R. and Larsen, M. (1983). *Neurosci. Lett.* **40**, 145–149.

41 Van Dijk, A.M.A., Ernst, A., and Schoenenberger, G.A. (1984). 8th Taniguchi Symp. Kyoto-Katata, Abstract, pp. 30–32.

42 Yehuda, S., Kastin, A.J., and Coy, D.H. (1980). *Pharmacol. Biochem. Behav.* **13**, 895–900.

43 Yehuda, S. and Mostofsky, D.I. (1982). *Int. J. Neurosci.* **16**, 221–226.

14

BIOCHEMICAL EVIDENCE FOR DSIP SPECIFIC BINDING SITES IN THE RAT BRAIN

ANDRÉ M.A. VAN DIJK AND GUIDO A. SCHOENENBERGER

Department of Surgery/Research Department, University Clinics, CH-4031 Basel, Switzerland

Delta-sleep-inducing peptide (DSIP: Trp-Ala-Gly-Gly-Asp-Ala-Ser-Gly-Glu) has been isolated, characterized, and sequenced from the extracorporeal dialysate of venous blood from the sinus confluens of rabbits during electrical stimulation of the interlaminar thalamus (*26*). Its name originates from the first discovered physiological action of the dialysate: an electroencephalogram (EEG) delta/spindles-inducing effect in recipient rabbits (*21*). Subsequently, evidence for multivariable central nervous system (CNS) interactions has been accumulated as well as effects of this peptide on the sleep pattern of various species (*10, 22, 23*). As a prerequisite for an efficient application, a bell-shaped dose response curve and an injection time dependence were experienced to be crucial characteristics of this peptide. For further details see the published reviews (*13, 21, 25*). Till now, a much wider activity spectrum for DSIP has been established. DSIP and its even more effective P-Ser[7]-DSIP (DSIP-P) analogue are able to influence and even to generate a circadian rhythm as measured by locomotor activity in the absence of the external "Zeitgeber," *i.e.*, in rats under continuous light conditions (*10, 11*). In addition, DSIP clearly affects the enzymatic activity of serotonin-N-acetyltransferase, a pineal enzyme known to exhibit strong circadian rhythmicity (14). Moreover, the peptide is able to influence daily brain levels of neurotransmitters and plasma protein levels (9).

The existence of DSIP and DSIP-P-like immunoreactive material (RIA) in several forms widely distributed throughout the brain, peripheral organs and body fluids has been firmly established (5, 18). Whereas in plasma mainly bound macromolecular forms were detected, in cerebrospinal fluid (CSF) around 50% is present in the free form and in urine predominantly the free form appeared to be excreted (5, 17). Recently it has been reported that RIA-like DSIP is also present in human breast milk (7, 12). By immunohistochemical studies DSIP could be detected in more restricted areas, particularly in the hypothalamus, hippocampus, thalamus, septum and brainstem nuclei of the rat (3, 4, 8). Variable amounts were found in the cortex and no immunoreactivity has been found in the cerebellum and the medulla (3). The only autoradiographical study was performed on neurons of cultured fetal rat brainstem (medulla-pons). Variable numbers of labeled cells were found from one culture to another. Binding was displaceable by excess unlabeled DSIP and localized predominantly in the neuronal cell bodies and primary dendrites, but not on glial cells (15).

I. PREVENTION OF DSIP BREAKDOWN IN A RAT BRAIN HOMOGENATE

In our laboratory, ^{3}H-DSIP has initially been used by intravenous injection into rats to determine its brain distribution pattern. However, the radioactive distribution pattern of ^{3}H-DSIP throughout the brain could not be distinguished from the pattern found after ^{3}H-Trp injections. The nonapeptide and its N-Tyr analogue pass intact and by a noncompetitive mechanism the blood-brain barrier (1). Breakdown studies revealed mainly N-terminal Trp liberation from DSIP by brain enzymes (16, 19), with a half-life time of about 15 min. Therefore clarification and effective inhibition of DSIP degradation became an unavoidable prerequisite before initiating any binding study.

Since the available ^{3}H-DSIP (synthesized by Dr. D. Gillessen, Hoffmann-La Roche) is exclusively labeled at the Trp residue, degradation to Trp after appropriate incubation time can be determined adequately by measuring the radioactive elution pattern after high performance liquid chromatography (HPLC) analyses. We developed several HPLC systems to separate not only Trp from DSIP (1-9), 5-hydroxytryptamine (5-HT) and 5OH-Trp, but also from DSIP (1-8), DSIP (2-9), DSIP (2-4) and DSIP (3-6). In HPLC system I 0.025% trifluoroacelic acid (TFA)/acetonitril (96:4 v/v) served as the mobile phase and elution from the deactivated Supelcosil LC-18-DB column was accomplished with 0.025% TFA/acetonitril (75:25 v/v) by means of a 1 hr hyperbolic gradient. In the HPLC system II a Brownlee Spherisorb 5 RP-18 was used with 1.0 M formic acid, pyridine (pH 2.9) as running buffer. Peptides were eluted by a 40 min isocratic, 1 hr linear gradient with 40% propanol in buffer A. The peptides were detected after postcolumn derivatization of primary amines by fluorescence at an excitation wavelength of 390 nm and an emission wavelength of 490 nm (2). A membrane preparation from whole rat brain served as the enzyme source. Briefly, freshly excised brains from male Wistar rats weighing 120–180 g were homogenized at 4°C in 20 volumes 0.32 M sucrose, 25 mM Tris, 5 mM $MgSO_4$ (HCL, pH 7.4). After differential centrifugation of 10 min 1,000 g and 20 min 27,000 g

TABLE I

Influence of Various Inhibitors on ^{3}H-DSIP Breakdown and Binding

Substance	Concentration	%$_0$ inhibition of ^{3}H-DSIP	
		Breakdown	Binding
EDTA	1.0 mM	14	40
Puromycin	0.1 mM	21	50
TRPHDM	0.1 mM	24	58
O-phenanthroline	0.5 mM	39	77
Bacitracin	200 μM	69	75
Substance P	30 μM	73	78
PCMBA	30 μM	92	79
PCMBA	100 μM	95	86
O-Phe/PCMBA	0.5 mM/30 μM	99	91
O-Phe/substance P	0.5 mM/30 μM	94	94
O-Phe/bacitracin	0.5 mM/200 μM	94	94

Incubations were performed with 3.2 nM ^{3}H-DSIP for 90 min at 22°C. Incubation without inhibitor served as 0 %$_0$ inhibition level whereas the control (0 min at 22°C) was taken as 100 % inhibition for ^{3}H-DSIP breakdown as well as ^{3}H-DSIP binding, which is defined as the difference in dpm/filter between incubations in the absence and in the presence of 10⁴ nM DSIP (1-9).

respectively (6), the membrane pellet was washed twice and finally homogenized 1:10 in 25 mM Tris, 5 mM MgSO$_4$ (HCL, pH 7.4). The enzymatic process was initiated by addition of 0.5 ml enzyme preparation (4 mg protein) to ^{3}H-DSIP (final concentration of 3.2 nM) with or without various peptidase inhibitors in a final volume of 2 ml. After 90 min at 22°C the incubation was stopped by 5 min boiling of the sample. In the subsequent HPLC-I analyses, 53% and 18% of the radioactivity were detected on the Trp and DSIP (1-8) position respectively, whereas only 26.4% still coeluted with DSIP (1-9). No radioactivity was found on 5-HT or 5OH-Trp positions, hence metabolic conversion of Trp played no significant role in our *in vitro* system. After control incubations of 0 min at 22°C, 90% radioactivity was recovered as DSIP (1-9), with up to 4% undefined radioactivity always located from fractions 4-8. The effective inhibitors for ^{3}H-DSIP breakdown in the CNS are summarized in Table I. EDTA, puromycin (PUR), and Trp-hydroxamate (TRPHDM) yielded only up to 24% inhibition, whereas O-phenanthroline, bacitracin, thiorphan, and substance P were moderately active (39–73%). Parachloromercuribenzoic acid (PCMBA) yielded up to 95% inhibition. Efficient prevention of ^{3}H-DSIP breakdown could be accomplished by combination of O-phenanthroline with either bacitracin, substance P or PCMBA (Table I). This suggests that ^{3}H-DSIP degradation in a rat brain homogenate involves thioendopeptidases (PCMBA), enkephalin-like aminopeptidases (substance P), and apparently metallopeptidases (O-phenanthroline).

II. DSIP BINDING IN THE PRESENCE OF VARIOUS INHIBITORS

Subsequently we attempted ^{3}H-DSIP binding in the same system by using a 1,000-fold higher concentration of DSIP (1-9) as displacer in the presence of the various peptidase inhibitors. Bound and free DSIP were separated by the Whatman GF/B filter technique and radioactivity was counted after elution from the glass filters by Quickzint: water (95:5 v/v).

However, addition of a more or less effective inhibitor reduced consequently ^{3}H-DSIP displaceable binding to rat brain membranes (Table I). Moreover, the impairment of ^{3}H-DSIP binding could be positively correlated to prevention of ^{3}H-DSIP breakdown after Line-

TABLE II

Effect of *In Vitro* Treatments on Binding of ³H-DSIP and ³H-Trp to Membranes from Rat Brain

Treatment		Binding (% of control)	
		³H-DSIP	³H-Trp
(A)			
Pre-incubation[a]			
30 min 22°C		89 ± 9	84 ± 7
37°C		55 ± 7	75 ± 8
Incubation[b]			
PCPA	(10 μM)	55 ± 5	52 ± 6
TRPHDM	(1 mM)	63 ± 6	76 ± 4
Mg^{2+}[c]	5 mM	132 ± 8	136 ± 12
Ca^{2+}[c]	5 mM	155 ± 37	175 ± 12
(B)			
O-phenanthroline[b]	(5 mM)	6 ± 5	84 ± 6
Bacitracin	(200 μM)	25 ± 6	86 ± 10
PCMBA	(30 μM)	21 ± 4	62 ± 5
NaCl[d]	130 mM	38 ± 9	74 ± 4
pH[e]	8.6	$28 + 9$	76 ± 10
	6.5	198 30	160 ± 8

³H-DSIP and ³H-Trp were defined as described in the text. Values, mean $\pm$ S.E.M. PCPA, parachlorophenylalanine. The respective controls were:

[a] Without pre-incubation.

[b] Without substance.

[c] With 1 mM of the appropriate ion.

[d] Without NaCl.

[e] At pH 7.4.

weaver-Burk plot analysis ($y = 0.21x + 9.46$; ρ: 0.98). This suggests an involvement of intrinsic enzymatic activity in ³H-DSIP binding, yielding ³H-Trp. Attempts to dissociate enzymatic activities from the DSIP binding site by water shock, multiple washings, acid treatment or freeze-thawing steps of the P$_2$ pellet, by further subcellular fractionation or by lower incubation temperature (4°C), were not successful. Hence, ³H-DSIP binding in the above described system seems to reflect merely ³H-Trp binding. Indeed changes in ³H-DSIP binding (13 nM displaced by 10^4 nM Trp) were highly correlated ($y = 0.99x + 1$; ρ: 0.96; Table II A). However, various enzyme inhibitors exerted a more pronounced inhibition on ³H-DSIP binding, a phenomenon also found by increasing NaCl concentrations and increasing pH; at lower pH both bindings increased to a similar extent (Table II B). From these data it may be concluded that even if ^{3}H-DSIP and ^{3}H-Trp binding may be closely related, they may not be identical. On the other hand, since the established discrepancies always revealed a diminished ^{3}H-DSIP binding and never an augmentation relative to ^{3}H-Trp binding, these results may be interpreted simply as an impairment of DSIP degradating activity due to the respective treatments.

III. DEVELOPMENT OF BINDING WITH THE DISPLACER DSIP (2-9)

A base for successful investigation of DSIP binding sites was provided by the possibility to eliminate enzymatic breakdown of the displacer to Trp by the use of DSIP (2-9). Intact DSIP (2-9) should only be able to displace intact ^{3}H-DSIP but not degradated label, *i.e.*, ^{3}H-Trp. Indeed, 10^4 nM DSIP (2-9) incubated for 3 hr at 10°C was able to displace ^{3}H-DSIP label for 22% (100% binding equalled 13,000 dpm/assay), whereas equimolar amounts of ^{3}H-Trp were only displaced for 2–5% by DSIP (2-9) from rat brain membranes. Compared to the displacement with 10^4 nM DSIP (1-9), the DSIP (2-9) displaceable ^{3}H-DSIP was even more affected by increasing NaCl concentrations (50% more inhibition by 20 to 160 mM NaCl) and bacitracin (from 20 to 200 μM, 50 to 30% more impairment). O-phenanthroline (0.1–1 mM) however, exerted a less pronounced inhibition on ^{3}H-DSIP displacement by DSIP (2-9) than by DSIP (1-9).

Subsequent optimization of the specific displaceable ^{3}H-DSIP binding by DSIP (2-9) resulted in the following experimental design, utilizing a membranal P_2 pellet according to Whittaker (27). Briefly, after homogenization of fresh rat brain minus cerebellum (since no immunoreactive DSIP has been found in this area (3), in ice-cold 0.32 M sucrose and centrifugation for 10 min 1,000 g, the supernatant was centrifuged for 55 min at 17,000 g. After three successive washings in incubation buffer of 25 mM morpholinopropanesulfonic acid (MOPS) and 1 mM MgSO$_4$ (KOH, pH 7.4) the resuspended P_2 pellet was incubated with 40 nM ^{3}H-DSIP (specific activity 2.1 Ci/mmol) with and without 10^5 nM DSIP (2-9) in a total volume of 2 ml (protein concentration 1 mg/ml).

Increasing magnesium concentrations from 1 to 11 mM not only resulted in a better displacement of ^{3}H-DSIP by DSIP (2-9), but at the

same time of ^{3}H-Trp. Since also ^{3}H-DSIP revealed a higher degree of
degradation (70% at 11 mM against 55% at 1 mM Mg^{2+}, by HPLC-II
analyses) these results are readily explained by increasing enzymatic
activity at higher magnesium concentrations. In the presence of 1 mM
Mg^{2+}, DSIP (2-9) was degraded for 23% after 5 hr incubation at 10°C
and subsequent HPLC-II analyses (Fig. 1). O-phenanthroline was able
to reduce this breakdown to 12%, whereas the aminopeptidase inhibitor

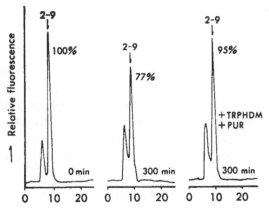

Fig. 1.   Fluorescence patterns after HPLC-II analysis of DSIP (2-9) after 0 and 300
min at 10°C, with and without 0.1 mM PUR/TRPHDM. Recoveries of DSIP (2–9) were
measured by peak heights.

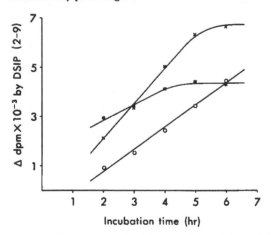

Fig. 2.   Temperature dependency of association of ^{3}H-DSIP to rat brain membranes.
Specific displaceable binding (Δdpm $\times 10^{-3}$) is defined as the difference between radioactivity bound in the absence and in the presence of 10^5 nM DSIP (2-9).   10°C; O
4°C; * 22°C.

PUR (0.1 mM) in combination with TRPHDM (0.1 mM, an enkephalin aminopeptidase inhibitor) rendered the DSIP (2-9) peptide fully intact (Fig. 1). Therefore, all subsequent experiments were performed in the presence of PUR/TRPHDM (0.1 mM) and specific DSIP binding was defined as the displacement of ³H-DSIP by 10⁵ nM DSIP (2-9).

Temperature dependencies for this binding revealed that a plateau was reached after 240 min at 22°C, after 300 min at 10°C and no equilibrium was reached up to 360 min at 4°C (Fig. 2). At equilibrium, specific DSIP binding appeared to be higher at 10°C than at 22°C, probably due to lower enzymatic activity. Subsequent incubations were performed at 10°C.

IV. SOME CHARACTERISTICS OF DSIP (2-9) DISPLACEABLE ³H-DSIP BINDING

After 300 min incubation at 10°C, the displacement curve of ³H-DSIP with DSIP (2-9) revealed a total and nonspecific binding of 12,000 and 5,000 dpm/assay respectively with an IC_{50} of 13.7 ± 0.6 μM (Fig. 3). The specific displaceable binding showed a linear protein dependency from 0.5–2.5 mg protein/assay. Scatchard analyses proved one binding site with a K_D of 15 μM and a B_{max} of 260 pmol/mg protein. The Hill

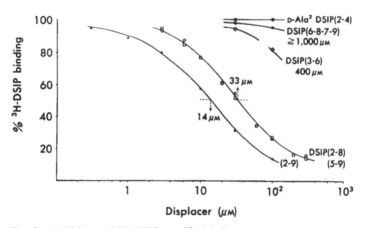

Fig. 3. Inhibition of ³H-DSIP specific binding to rat brain membranes by DSIP fragments. Membranes were incubated with 40 nM ³H-DSIP and varying concentrations of peptides for 5 hr at 10°C. The points are means of triplicate determinations differing less than 10 %. The IC_{50} values are indicated in μM.

coefficient was close to unity ($n=1.02$), also indicating a pure 1:1 "receptor"-ligand interaction.

Several smaller DSIP fragments were tested in displacing ^{3}H-DSIP and their affinities compared to the one of DSIP (2-9) (Fig. 3). Shortening from the C-terminal site with one amino acid resulting in DSIP (2-8) reduced the affinity about 3-fold. The fragment D-Ala[2]-DSIP (2-4) was completely inactive. Hence the DSIP (2-9) interaction site may be mainly situated in the C-terminal part of the molecule. Accordingly, shortening of the peptide from the N-terminal site to DSIP (5-9) revealed the same potency as DSIP (2-8), whereas DSIP (6-8-7-9) showed negligible displacement. It is unlikely that the inversion of positions 7 and 8 would account for the complete loss in activity. Hence position 5 (Asp) seems to play a crucial role in the interaction, although more C-terminal amino acids are needed for full expression since the fragment DSIP (3-6) revealed an almost 30 times reduced affinity. It is conceivable that more or less preservation of the proposed DSIP ring structure (20) in which position 5 plays a central role, determines the affinity of DSIP (2-9) to the DSIP binding site.

Addition of increasing NaCl concentrations to the incubation medium resulted in competitive inhibition of DSIP (2-9) displaceable

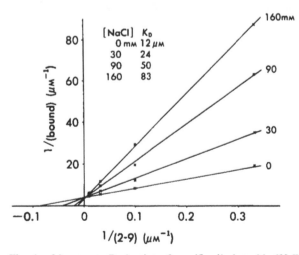

Fig. 4. Lineweaver-Burk plot of specific displaceable ^{3}H-DSIP by DSIP (2-9) and its inhibition by NaCl at the concentrations indicated. Slopes of the lines were determined by the method of the least squares. The K_D's were calculated from the intercept with the X-axis.

^{3}H-DSIP binding as calculated from the Lineweaver-Burk plot (Fig. 4). Up to a 7-fold reduction in DSIP (2-9) affinity occurred in the presence of 160 mM NaCl without a change in the B_{max} (250 pmol/mg protein).

A primary estimation of the brain distribution pattern of ^{3}H-DSIP binding has been obtained by dissecting the cerebral cortex, cerebellum, medulla-pons, and the remaining part of the brain. Taking displaceable ^{3}H-DSIP/mg protein of whole brain as 100%, highest values of DSIP (2-9) displaceable binding were found in the cortex (115%) and rest of the brain (71%) and significantly less in cerebellum (38%) and medulla-pons (22%). The lower values in the last two areas are in accord with the immunohistochemical data (3), where no immunoreactivity has been detected in cerebellum and medulla.

V. COMPARISON OF DSIP (2-9) AND DSIP (1-9) DISPLACEMENT

In Table III the various K_D's as determined from displacement curves for Trp, DSIP (1-9) and DSIP (2-9) are summarized. Although the affinity of Trp for ^{3}H-DSIP equals the one for ^{3}H-Trp, the maximal number of binding sites as determined by Scatchard analyses is 270 pmol/mg protein and 800 pmol/mg protein for ^{3}H-DSIP and ^{3}H-Trp displaceable binding, respectively. Therefore the two binding sites are not identical.

The DSIP (1-9) displacement curve not only exerts an 80-fold higher affinity to ^{3}H-DSIP as compared to ^{3}H-Trp binding, but gives a biphasic Scatchard plot with K_{D1} of 5.7 μM and a B_{max} of 110 pmol/mg protein and a K_{D2} of 24.7 μM and a B_{max} of 230 pmol/mg protein. The second K_D and B_{max} are in the same range as the ones for the degradation products of DSIP (1-9), i.e., DSIP (2-9) and Trp. Indeed, DSIP (1-9) is still degradated for about 50% as judged by HPLC-II analyses. In combination with the finding that DSIP (1-9) and not

TABLE III
Affinities to DSIP and Trp binding Sites

Displacer	Ligand (μM)	
	^{3}H-DSIP	^{3}H-Trp
Trp	14.0	14.5
DSIP (1-9)	3.5	275
DSIP (2-9)	13.7	1,000

DSIP (2-9) is able to displace ^{3}H-Trp in the presence of PUR/TRPHDM, the second K_D and B_{max} may solely be due to the generation of Trp. Hence, Trp may act through its own as well as the DSIP binding site, which may provide an explanation for the sleep effects of Trp *in vivo* (*24*).

It is conceivable that the generation of Trp during incubation shifts the affinity of DSIP (1-9) to its binding site from the nM to the μM range. Therefore, until the presently described system is not optimalized with respect to intact DSIP (1-9), no conclusion may be drawn concerning the true affinity of DSIP (1-9) to its receptor.

In conclusion, the present study provides convincing biochemical evidence for binding of intact ^{3}H-DSIP (1-9) and its specific displacement in the CNS of the rat. Since ^{3}H-Trp could neither be displaced by DSIP (2-9) nor by the smaller DSIP fragments tested, the characteristics of the DSIP (2-9) displacement support the hypothesis for the existence of specific DSIP binding sites in the CNS.

SUMMARY

A specific binding assay for DSIP in the rat brain has been developed. After 90 min incubation at 22°C with rat brain homogenate the peptide was only intact for 26%. Efficient prevention of breakdown occurred in the presence of O-phenanthroline with either bacitracin, substance P or PCMBA. However, no measurable DSIP binding could be detected in the presence of these inhibitors. By using DSIP (2-9) as displacer instead of DSIP (1-9), the system could be optimalized for specific displaceable ^{3}H-DSIP binding at lower incubation temperature (10°C) in the presence of PUR/TRPHDM. After 5 hr equilibrium was reached and DSIP (2-9) could displace ^{3}H-DSIP from its binding site with a K_D of 14 μM. Structure affinity studies with smaller fragments indicate an interaction site at the C-terminal part (5-9) of DSIP (2-9). Temperature dependency, NaCl dependency and a primary brain distribution pattern have been characterized. Although DSIP (1-9) is still degradated for about 50% in this system, the affinity of DSIP (1-9) for the DSIP binding site is already 80-fold higher than for the Trp binding site. Scatchard analyses revealed a different maximal number of binding sites for DSIP and Trp. In conclusion, the DSIP and Trp binding sites are not identical, and enough intact ³H-DSIP binds to rat membranal tissue to be specifically displaced by DSIP (2-9).

REFERENCES

1 Banks, W.A., Kastin, A.J., and Coy, D.H. (1984). *Brain Res.* **301**, 201–207.
2 Böhlen, P., Stein, S., Stone, J., and Udenfriend, S. (1975). *Anal. Biochem.* **67**, 438–445.
3 Constantinidis, J., Bouras, C., Guntern, R., Taban, C.H., and Tissot, R. (1983). *Neuropsychobiology* **10**, 94–100.
4 Constantinidis, J., Bouras, C., and Richard, J. (1983). *Clin. Neuropathol.* **2**, 47–54.
5 Ekman, R., Larsson, I., Malmquist, M., and Thorell, J.I. (1983). *Regul. Peptides* **6**, 371–378.
6 Emson, P.C., Lee, C.M., and Rehfeld, Y.F. (1980). *Life Sci.* **26**, 2157–2163.
7 Ernst, A., Monti, J.C., and Schoenenberger, G.A. In *Sleep 1984*, ed. Koella, W.P., Ruether, R., and Schulz, H. Stuttgart, New York: Gustav Fischer Verlag. (in press).
8 Feldman, S.C. and Kastin, A.J. (1984). *Neuroscience* **11**, 303–307.
9 Graf, M., Baumann, J.B., Girard, J., Tobler, H.J., and Schoenenberger, G.A. (1982). *Pharmacol. Biochem. Behav.* **17**, 511–517.
10 Graf, M., Christen, H., and Schoenenberger, G.A. (1982). *Peptides* **3**, 623–626.
11 Graf, M., Christen, H., Tobler, H.J., Maier, P.F., and Schoenenberger, G.A. (1981). *Pharmacol. Biochem. Behav.* **15**, 717–721.
12 Graf, M.V., Hunter, C.A., and Kastin, A.J. (1984). *J. Clin. Endocrinol. Metab.* **59**, 127–132.
13 Graf, M.V. and Kastin, A.J. (1984). *Neurosci. Biobehav. Rev.* **8**, 83–93.
14 Graf, M.V., Kastin, A.J., and Schoenenberger, G.A. (1985). *J. Neurochem.* **44**, 629–632.
15 Hösli, E., Schoenenberger, G.A., and Hösli, L. (1983). *Brain Res.* **279**, 374–376.
16 Huang, J.-T. and Lajtha, A. (1978). *Res. Commun. Chem. Pathol. Pharmacol.* **19**, 191–199.
17 Kastin, A.J., Castellanos. P.F., Banks, W.A., and Coy, D.H. (1981). *Pharmacol. Biochem. Behav.* **15**, 969–974.
18 Kastin, A.J., Nissen, C., Schally, A.V., and Coy, D.H. (1978). *Brain Res. Bull.* **3**, 691–695.
19 Marks, N., Stern, F., Kastin, A.J., and Coy, D.H. (1977). *Brain Res. Bull.* **2**, 491–493.
20 Mikhaleva, I., Sargsyan, A., Balashova, T., and Ivanov, V. (1982). In *Chemistry of Peptides and Proteins*, ed. Voelter, W., Wuensch, E., Ovchinnikov, J., and Ivanov, V., vol. 1, pp. 289–297. Berlin: de Gruyter.
21 Monnier, M. and Schoenenberger, G.A. (1983). In *Functions of the Central Nervous System*, ed. Monnier, M. and Meulders, M., vol. 4, pp. 161–219. Amsterdam, New York, Oxford: Elsevier.
22 Nagasaki, H., Kitahama, K., Valatx, J.L., and Jouvet, M. (1980). *Brain Res.* **192**, 276–280.
23 Polc, P., Schneeberger, J., and Haefely, W. (1978). *Neurosci. Lett.* **9**, 33–36.
24 Schneider-Helmert, D. (1981). *Int. Pharmacopsych.* **16**, 162–178.
25 Schoenenberger, G.A. (1984). *Eur. Neurol.* **23**, 321–345.
26 Schoenenberger, G.A. and Monnier, M. (1977). *Proc. Natl. Acad. Sci. U.S.* **74**, 1282–1286.
27 Whittaker, V.P. (1969). In *Handbook of Neurochemistry*, ed. Lajtha, A., vol. 2, pp. 237–251. New York: Plenum Press.

MURAMYL PEPTIDES, PROSTAGLANDINS, AND SLEEP-PROMOTING SUBSTANCE

15

MURAMYL PEPTIDES AND INTERLEUKIN-1
AS PROMOTERS OF SLOW WAVE SLEEP

JAMES M. KRUEGER

Department of Physiology and Biophysics, University of Health Sciences/The Chicago Medical School, North Chicago, Illinois 60064, U.S.A.

The ancient observation that prolonged wakefulness increases the desire to sleep led to the hypothesis that during wakefulness a sleep-promoting substance(s) (hypnotoxin, neuromodulator) accumulates in brain. Piéron and colleagues were the first to experimentally examine this hypothesis; they described the presence of a thermolabile nondialyzable substance in cerebrospinal fluid (CSF) of sleep-deprived dogs (*31*). Since that time numerous laboratories have described a variety of sleep-promoting substances (*17*); these include arginine vasotocin (*39*), delta-sleep-inducing peptide (DSIP) (*41*), uridine (*16*), prostaglandin D_2 (PGD_2) (*44*), and factor S (*10*). Each of these substances has unique chemical and biological characteristics. The somnogenic actions of the latter substance, factor S, and related substances are the focus of this review.

The accumulation of factor S in CSF during wakefulness was first described by John Pappenheimer and colleagues (*10*). Although the biological actions of factor S derived from CSF were described and certain of its chemical properties were determined by Pappenheimer (*10, 36, 37*), insufficient amounts of CSF were available for complete chemical characterization of factor S. Brain (*37*) and urine (*20*) were used, there-

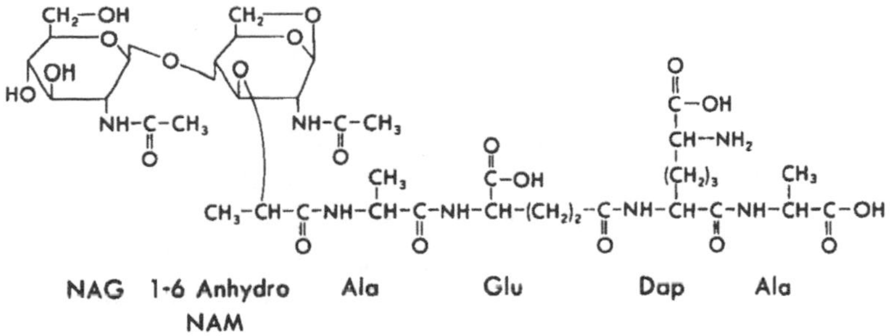

Fig. 1. Urinary factor S (FSu) (*32*). Other muramyl peptides (MPs), synthetic and natural, were also active as somnogens as well as being pyrogenic, immunoadjuvants, and promoters of non-specific host resistance, depending on structure (*23*). Results from structure-sleep promoting activity studies suggest: 1) the 1-6 anhydro ring on NAM, 2) stereospecificity (*22*), 3) amidation-deamidation of key carboxyl groups (*23*), and 4) the presence of NAM but not NAG (*23*), are important determinants of biological activity. One pmol of FSu was sufficient to induce excess SWS for several hours (*25*) (see Fig. 2). Other MPs are less potent, *e.g.*, 50 pmol of MDP was needed to induce results similar to those induced by 1 pmol of FSu.

fore, to permit large scale extractions of a similar, perhaps identical, substance. The sleep-promoting materials derived from these sources were identified as muramyl peptides (MPs) (*21*). The urinary material (FSu) was completely characterized; it is N-acetylglucosaminyl-1-6-anhydro-N-acetylmuramyl-alanyl-glutamyl-diaminopimelyl-alanine (NAG-1-6-anhydro-NAM-Ala-Glu-Dap-Ala) (Fig. 1) (*25, 32*).

MPs are common constituents of peptidoglycans of bacterial cell walls. Many naturally occurring and synthetic MPs such as NAM-Ala-Isogln (MDP) have immunomodulatory and pyrogenic activities (*30*). Some of these MPs also have sleep-promoting effects similar to those elicited by FSu (*22, 23*). The immunologic and febrile effects of MPs are thought to be mediated through the leukocytic monokine interleukin-1 (IL1) (*29*) (also called endogenous pyrogen). This consideration led us to assay IL1 for sleep-promoting activity. Purified preparations of IL1 induced dose-dependent increases in slow wave sleep (SWS). These recent experimental advances as well as hypotheses that developed from them are discussed below.

I. SLEEP-PROMOTING ACTIVITY OF FSu AND OTHER MPs

The actions of various somnogenic MPs are similar although the dose of each needed to elicit sleep effects varies. Most information concerning somnogenic actions of MPs was obtained using rabbits as the assay animal (*21–23*). Rabbits exhibit short bouts of SWS which are sometimes followed by REM sleep episodes. Typically, SWS bouts occupy about 40±7% of daylight hours (*38*). Following intracerebral ventricular (i.c.v.) administration of somnogenic MPs, duration of sleep often increased to 60–70%. During the first hour post-administration duration of SWS was usually similar to control values. Excess SWS became evident during the second hour and then persisted for 6 or more hours (Fig. 2). The episodic nature of rabbit sleep persisted during periods of excess SWS induced by MPs. Increased duration of SWS was primarily the result of an increase in the number of sleep episodes lasting 4 or

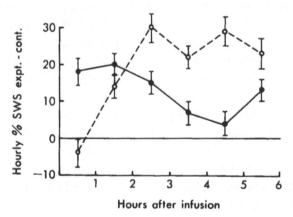

Fig. 2. Time courses of effects of FSu (O) and IL1 (●) on rabbit SWS. Samples in 0.3 ml of artificial CSF were infused at a rate of 7 μl/min; EEG, activity, and brain temperature were then recorded over the next 6 hr. Each rabbit was used as his own control. Values shown represent the means±S.E.M. of individual experimental-control differences. Five μl of purified human IL1 (*n*=9) and 1–5 pmol of FSu (*n*=11) were infused. Following IL1 infusions, excess SWS was observed during the first post-infusion hour; with FSu excess SWS was not evident until the second hour post-infusions. The time course of sleep effects elicited by other somnogenic MPs was similar to that of FSu. There were also differences in the time courses of fever responses (not shown). Immediately after IL1 infusions, animals were febrile. Following FSu fevers developed more slowly.

more min (20). The effects of MPs on rabbit REM sleep have not been systematically studied. However, preliminary results from our laboratory and those of Scherschlicht and Marias (40) suggest that MPs inhibit REM sleep in rabbits. These results are similar to those obtained from cats, rats, and monkeys (see below). Similarly, an inhibition of REM during the first night of recovery is observed in humans following prolonged sleep-deprivation (14).

Somnogenic MPs also induced increases in amplitudes of electroencephalogram (EEG) slow waves (1/2–4 Hz) during bouts of SWS (21, 22). Similar increases in slow wave amplitudes were observed during the deep sleep that followed sleep-deprivation in rabbits (37), rats (2), and man (3). Other aspects of sleep-waking behavior following MP administration also appeared normal. As mentioned above, the episodic nature of sleep was retained, sleep postures remained normal, and animals could be aroused easily by noise or other disturbances. Typical spontaneous waking behaviors such as eating, drinking, and grooming were also observed following administration of MPs. However, following high doses of FSu or other MPs abnormalities such as excess nasal secretions (22) were observed in rabbits.

The effects of MPs on cat sleep were similar to those observed in rabbits. Administration of FSu at the beginning of daylight hours induced prolonged increases in excess SWS although this effect was not observed until 3–4 hr after infusion (13). During these hours, REM sleep was inhibited. Latency to SWS was reduced following intentional awakings during periods of maximum excess SWS (4–10 hr post-infusion). During this period normal cat SWS-REM-waking cycles were observed. The day after FSu administration, durations of SWS and REM were similar to control values.

In rats the effects of MPs appear more complex. Rats exhibit large circadian variations in the amount of time spent in sleep. During dark hours rats are active whereas during daylight hours they typically sleep 70–80% of the time. Several laboratories (18, 20, 33) reported that excess SWS was induced by administration of MPs at the onset of dark hours. However, SWS was not significantly increased after administration of MDP at the onset of daylight hours (12, 33).

In squirrel monkeys sleep responses induced by MDP were also dependent upon the time of day injections were made (47). Sleep in this species is maximum during dark hours. If MDP was given early in the subjective day duration of SWS increased from control values of about 13% to over 50%. However, administration during the circadian night time induced only small increases in SWS. During the day squirrel monkeys normally do not exhibit REM sleep and MDP administered at this time failed to induce REM. REM sleep was inhibited following night time administration of MDP.

In summary the following characteristics of sleep induced by MPs have emerged. 1) MPs induce an increase in duration of SWS. 2) MPs induce an increase in EEG slow wave amplitudes during SWS. 3) MPs appear to transiently inhibit REM sleep. 4) Specific responses induced by MPs are dependent upon the time of administration. 5) Sleep cycles remain intact and the "drive" for sleep is greater (as measured by reduced latencies to SWS) following administration of MPs. 6) Sleep-waking behaviors are normal following low doses of MPs. Similar parameters characterize the sleep-waking patterns that follow sleep deprivation.

II. IL1: SOMNOGENIC ACTIONS

IL1 is synthesized and released by macrophages (9). IL1 has also been described as a product of central nervous system (CNS) astrocytes (11). Basal rates of IL1 release are low in both macrophages and astrocytes. Both cell types increase synthesis and release of IL1 in response to various bacterial products. Although the structure of IL1 remains unknown, the recent demonstration of an interaction between IL1 and a monoclonal anti-MDP antibody suggests that IL1 itself may contain an MDP-like moiety (7). We found that IL1 was capable of inducing increases in SWS (24). These findings led us to speculate that IL1 may be part of the series of biochemical events that are responsible for MP induced sleep.

In our first set of experiments we compared the sleep-promoting effects of dialyzed supernates of rabbit macrophages stimulated by a MP to those elicited by dialyzed supernates from control cells. Only the supernates obtained from stimulated cells induced significant increases in SWS following i.c.v. infusion. In later experiments, purified human IL1 obtained from heat-killed *Staphylococcus albus* stimulated leukocytes was used. i.c.v. infusion of these preparations also induced dose-dependent increases in SWS. Mild heat treatment inactivates IL1 and this treatment also destroyed sleep-promoting activity. Thus IL1 resembles the sleep-promoting material described by Piéron in that it is heat labile and nondialyzable. Our work with IL1 was confirmed by Tobler and colleagues (43) who reported that murine IL1 derived from astrocytes induced increased slow-wave activity in rats. Moldofsky and colleagues (34) found that human IL1 blood concentrations peaked at the onset of SWS. Other indirect evidence that IL1 may be involved in sleep regulation stems from postmarathon contestants. These individuals exhibited increase in circulating leukocytes and IL1-like activity (6). In another study, similar contestants exhibited excess SWS for several nights following marathon runs (42).

In rabbits, the excess SWS following IL1-i.c.v. infusions was evident during the first hour post-infusion. In contrast, sleep effects following FSu-i.c.v. infusions were not observed until the second hour after infusion (Fig. 2). Similarly, following intravenous administration of IL1 excess SWS was observed during the first hour after injections whereas excess SWS was not observed until the second hour after MDP intravenous injections. These time courses suggest that the somnogenic actions of MPs may be mediated through a step involving synthesis and release of IL1.

IL1 preparations also induced increases in EEG slow-wave amplitudes similar to those observed following FSu infusion or sleep deprivation as described above. IL1 also induced a transient inhibition of REM sleep during those periods when the greatest amount of excess SWS was observed (unpublished data). Other aspects of sleep-waking behaviors following IL1 administration appeared normal as described above for MP treated animals and post-sleep-deprived animals.

III. EFFECTS OF MPs AND IL1 ON BODY TEMPERATURE

It is well known that normal sleep is usually accompanied by a decrease in body temperature rather than an increase. However, MPs and IL1 are pyrogenic. This may, at first, seem inconsistent with the hypothesis that these substances induce sleep through normal mechanisms. However, we have shown that fever and sleep responses induced by IL1 and

MPs could be separated from each other. Further, the normal temperature changes associated with the transition between wakefulness-SWS-REM persisted in MP and IL1 treated animals.

The pyrogenic actions of IL1 could be blocked without affecting sleep-responses when an antipyretic, anisomycin, was co-infused with IL1 (24). Similarly, febrile responses but not sleep responses induced by MDP were attenuated after acetaminophen pretreatment of animals (22). Certain MPs were found to be pyrogenic but not somnogenic (23). Thus sleep responses do not appear to be secondary to fever responses although it is possible that certain biochemical events involved in each response may be shared.

In rabbits the transition between wakefulness and SWS is associated with a regulated decrease in body temperature (15). In contrast, if rabbits are acclimated and housed at 23°C, there is a relatively rapid increase in body temperature following the onset of REM sleep. These sleep related temperature changes persist in IL1 (26) and MP (unpublished data) treated animals even though they are febrile. These results suggest that those mechanisms responsible for fever induction are separate from the thermoregulatory mechanisms that are tightly coupled to sleep states.

IV. HYPOTHESES

Many relationships between temperature regulation and sleep have been described. Under certain circumstances a number of putative sleep neuromodulators (e.g., DSIP (48), PGD$_2$ (Fig. 3), and vasoactive intestinal peptide (VIP) (8)) also induce increases in body temperature. Further, many neurotransmitters thought to be involved in sleep regulation are also thought to be involved in thermoregulation (e.g., serotonin, dopamine, acetylcholine, etc.) (4). Nevertheless, normal sleep is not accompanied by fever and we must, therefore, explain why IL1 and MP simultaneously induced sleep and fever responses. The method by which we administered these substances was rather unphysiological in the sense that relatively large amounts of exogenous material were rapidly delivered to the entire brain. It is highly likely that under normal circumstances sleep and/or fever inducing substances would be delivered in a stringently regulated fashion to discrete areas which would provide

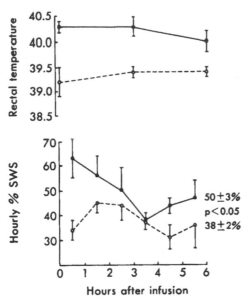

Fig. 3. Effects of PGD₂ on rabbit SWS and rectal temperatures. Samples were assayed as described in the legend to Fig. 2. Ueno and colleagues reported that PGD₂ induced increased sleep and hypothermia in rats (*44*). Prostaglandins are of interest because alterations in prostaglandin metabolism are thought to be involved in IL1 induction of fever and immune responses (*9*). In our rabbit assay system 25 pmol of PGD₂ failed to affect either sleep or temperature (results not shown). However, after infusion of 250 pmol of PGD₂ (●), excess SWS and increases in body temperature were induced. The time course of PGD₂ effects was similar to that observed following IL1 infusions but not that following MP administration (see Fig. 2). In rats PGD₂ induced hypothermia whereas in rabbits hyperthermia was induced; in both species excess SWS was induced. This reinforces the postulate that some aspects of temperature regulation are separate from sleep; see text for a discussion of this matter.

for very specific responses. Indeed, preliminary results from our laboratory (*46*) suggest that unilateral microinjection of IL1 into the preoptic anterior hypothalamus induces fever but not excess SWS. It is also possible that other temperature and sleep-wake modulators are released in concert with MPs and/or IL1 thereby creating normal sleep-waking-temperature cycles.

The close relationship between homeothermic temperature regulation and SWS is of interest since it appears that both evolved simultaneously. IL1 is found in lower vertebrates and it can induce behavioraly regulated fever in lizards (*19*). It is interesting to speculate that IL1, which is phylogenetically ancient and/or the even older MPs, may have been gradually incorporated as autonomic modulators of body temperature and SWS during the evolution of homeotherms. Co-evolution of SWS with warmbloodedness may have minimized the caloric cost of the latter. Thus it may be reasonable that similar and/or identical substances are involved in the regulation of both sleep and temperature.

We are proposing that one of the highest functions of man and other mammals, consciousness, is controlled, in part, by substances derived from unicellular organisms, MPs. Our work and that of others raise the question: What role do MPs play in mammalian physiology? There are no known mammalin synthetic pathways for some of the constituents of MPs, e.g., muramic acid and diaminopimelic acid. However, MPs have many biological activities in mammals; these include, modulation of sleep, temperature, and immune regulation and inhibition of angiotensin converting enzyme (5). Muramic acid is found in hydrolyzable linkage in mammalian tissue (49) and dap is a normal constituent of mammalian urine (28). Further, macrophages release MPs following uptake and digestion of bacteria (45). Another likely source of MPs is from gut, resulting from the breakdown of bacterial flora. Although precise quantitation of active MP levels in mammalian tissue has not been determined, Zhai Sen and Karnovsky (49) estimated that muramic acid (found in hydrolyzable linkage) levels in liver were about 150 pmol/g. A single pmol of FSu was sufficient to greatly alter sleep and temperature. These considerations have led to the postulate that muramyl peptides may be vitamin-like, i.e., that they are substances required by but cannot be synthesized by the host (1).

Amid extensive investigation and speculation the functions of sleep remain unknown. We have shown that two classes of immunologically active compounds, MPs and IL1, are also somnogenic. Many areas in brain involved in regulation of sleep are also known to be involved in CNS regulation of immune responses (27). Further, sleep deprivation has many effects on the immune system (35) as well as sleep. These considerations led us to postulate that sleep serves an immune function (27). That is, sleep may be an adjunct to the immune system that aids in the recovery from environmental challenges encountered during waking activity. Such a relationship is consistent with the "folk wisdom" that bed rest aids in prevention of any recovery from pathological states.

SUMMARY

Sleep-promoting Factor S derived from urine (FSu) was characterized as NAG-1-6-anhydro-NAM-Ala-Glu-Dap-Ala. Other MPs were somnogenic in rabbits, rats, cats, and monkeys. Sleep induced by MPs had the following characteristics. 1) MPs induced increases in duration of SWS. 2) MPs induced increases in amplitudes of EEG slow-waves. 3) MPs transiently inhibited REM. 4) Sleep responses elicited by MPs were dependent upon the time of day of MDP administration. 5) Sleep cycles following MP administration remained intact. 6) Sleep-waking behaviors were normal following low doses of MPs. Similar sleep-waking patterns were observed following sleep deprivation.

ILl preparations also induced dose-dependent increases in SWS concomitant with increasing body temperature. An antipyretic blocked ILl induced fever without affecting its sleep-promoting activity. Temperature changes tightly coupled to sleep states persisted in ILl treated animals. The time courses of MP and ILl induced sleep are consistent with the proposal that MPs induce sleep through a step involving ILl.

Acknowledgments

The author is grateful to Drs. Pappenheimer, Karnovsky, Chedid, and Dinarello who contributed to the results and ideas presented here. I also thank Dr. James Walter who has allowed me to mention his results that are not yet published in full form. This work was supported by the Office of Naval Research, U.S.A., contract No. N00014-82-K-0393 and by the National Institutes of Health, No. GM33268.

REFERENCES

1 Adam, A. and Lederer, E. (1984). *Med. Res. Rev.* **4**, 111–152.
2 Borbély, A.A. and Neuhaus, H.U. (1979). *J. Comp. Physiol.* **133**, 71–87.
3 Borbély, A.A., Baumann, R., Brandeis, D., Strauch, I., and Lehmann, D. (1981). *Electroenceph. Clin. Neurophysiol.* **51**, 483–493.
4 Blatteis, C.M. (1981). *Fed. Proc.* **40**, 2735–2740.
5 Bush, K., Henry, P.R., and Slusarchyk, J. (1984). *J. Antibiot.* **37**, 330–335.
6 Cannon, J.G. and Kluger, M.J. (1983). *Science* **220**, 617–619.
7 Chedid, L., Bahr, G.M., Riveau, G., and Krueger, J.M. (1984). *Proc. Natl. Acad. Sci. U.S.* **81**, 5888–5891.

8 Clark, W.G., Lipton, J.M., and Said, S.I. (1978). *Pharmacology* 17, 883–885.

9 Dinarello, C.A. (1984). *Rev. Infect. Dis.* 6, 51–95.

10 Fencl, V., Koski, G., and Pappenheimer, J.R. (1971). *J. Physiol.* 216, 565–589.

11 Fontana, A., Kristensen, F., Dubs, R., Gemsa, D., and Weber, E. (1982). *J. Immunol.* 129, 2413–2419.

12 Fornal, C., Karkus, R., and Radulovacki, M. (1984). *Peptides* 5, 91–95.

13 Garcia-Arraras, J.E. (1981). *Am. J. Physiol.* 241, E269–E274.

14 Hauri, P. (1982). *The Sleep Disorders, Current Concepts.* Kalamazoo: Upjohn Company.

15 Heller, H.C. and Glotzbach, S.F. (1977). *Int. Rev. Physiol.* 15, 147–187.

16 Honda, K., Komoda, Y., Nishida, S., Nagasaki, H., Higashi, A., Uchizono, K., and Inoué, S. (1984). *Neurosci. Res.* 1, 243–252.

17 Inoué, S., Uchizono, K., and Nagasaki, H. (1982). *Trends Neurosci.* 5, 218–220.

18 Inoué, S., Honda, K., Komoda, Y., Uchizono, K., Ueno, R., and Hayaishi, O. (1984). *Proc. Natl. Acad. Sci. U.S.* 81, 6240–6244.

19 Kluger, M.J. (1979). *Fed. Proc.* 38, 30–34.

20 Krueger, J.M., Bacsik, J., and Garcia-Arraras, J.E. (1980). *Am. J. Physiol.* 238, E116–E123.

21 Krueger, J.M., Pappenheimer, J.R., and Karnovsky, M.L. (1982). *J. Biol. Chem.* 257, 1664–1669.

22 Krueger, J.M., Pappenheimer, J.R., and Karnovsky, M.L. (1982). *Proc. Natl. Acad. Sci. U.S.* 79, 6102–6106.

23 Krueger, J.M., Walter, J., Karnovsky, M.L., Chedid. L., Choay, J., LeFrancier, P., and Lederer, E. (1984). *J. Expt. Med.* 159, 68–76.

24 Krueger, J.M., Walter, J., Dinarello, C.A., Wolff, S.M., and Chedid, L. (1984). *Am. J. Physiol.* 246, R994–R999.

25 Krueger, J.M., Karnovsky, M.L., Martin, S.A., Walter, J., Pappenheimer, J.R., and Biemann, K. (1984). *J. Biol. Chem.* 259, 12659–12662.

26 Krueger, J.M. and Walter, J. (1984). *Soc. Neurosci. Abstr.* 10, 505.

27 Krueger, J.M., Walter, J., and Levin, C. (1985). In *Brain Mechanism of Sleep*, ed. McGinty, D., pp. 253–275. New York: Raven Press.

28 Krysciak, J. (1980). *Folia Biol.* (Krakow) 28, 47–51.

29 Leclerc, C. and Chedid, L. (1982). *Lymphokines* 7, 1–21.

30 Lederer, E. (1980). *J. Med. Chem.* 23, 819–825.

31 Legendre, R. and Piéron, H. (1913). *Z. Allg. Physiol.* 14, 235–262.

32 Martin, S., Karnovsky, M.L., Krueger, J.M., Pappenheimer, J.R., and Biemann, K. (1984). *J. Biol. Chem.* 259, 12652–12658.

33 Masek, K., Kadlecova, O., Kadlec, O., Zidek, Z., Farghalli, H., and Machkova, Z. (1985). In *Muramyl Dipeptides.* New York: Bioscience Ediprint Inc. (in press).

34 Moldofsky, H., Gorczynski, R.M., Lue, F.A., and Keystone, E. (1984). *Sleep Res.* 13, 42.

35 Palmblad, J., Petrini, B., Wasserman, J., and Akerstedt, T. (1979). *Psychosomatic Med.* 41, 273–278.

36 Pappenheimer, J.R., Miller, J.B., and Goodrich, V.A. (1967). *Proc. Natl. Acad. Sci. U.S.* 58, 513–518.

37 Pappenheimer, J.R., Koski, G., Fencl, V., Karnovsky, M.L., and Krueger, J.M. (1975). *J. Neurophysiol.* 38, 1299–1311.

38 Pappenheimer, J.R. (1983). *J. Physiol.* 336, 1–11.

39 Pavel, S., Psatta, P., and Goldstein, R. (1977). *Brain Res. Bull.* 2, 251–254.

40 Scherschlicht, R. and Marias, J. (1983). *Experientia* **39**, 683.

41 Schoenenberger, G.A. and Monnier, M. (1977). *Proc. Natl. Acad. Sci. U.S.* **74**, 1282–1286.

42 Shapiro, C.M., Bortz, R., Mitchell, D., Bartel, P., and Jooste, P. (1981). *Science* **214**, 1253–1254.

43 Tobler, I., Borbély, A.A., Schwyzer, M., and Fontana, A. (1984). *Eur. J. Pharmacol.* **104**, 191–192.

44 Ueno, R., Honda, K., Inoué, S., and Hayaishi, O. (1983). *Proc. Natl. Acad. Sci. U.S.* **80**, 1735–1737.

45 Vermeulen, M.W. and Grey, F.R. (1983). *Fed. Proc.* **42**, 1221.

46 Walter, J., Krueger, J.M., Meyers, P., and Dinarello, C.A. (1984). *Soc. Neurosci. Abstr.* **10**, 505.

47 Wexler, D.B. and Moore-Ede, M.C. (1984). *Am. J. Physiol.* **247**, R672–R680.

48 Yehuda, S., Kastin, A. and Coy, D. (1980). *Pharmacol. Biochem. Behav.* **13**, 895–900.

49 Zhai Sen and Karnovsky, M.L. (1984). *Infect. Immun.* **46**, 937–941.

16

PROSTAGLANDIN D_2 REGULATES PHYSIOLOGICAL SLEEP

RYUJI UENO,[*1] OSAMU HAYAISHI,[*1] HIROYOSHI OSAMA,[*1] KAZUKI HONDA,[*2] SHOJIRO INOUÉ,[*2] YOUZOU ISHIKAWA,[*3] AND TERUO NAKAYAMA[*3]

*Hayaishi Bioinformation Transfer Project, Research Development Corporation of Japan,[*1] Institute for Medical and Dental Engineering, Tokyo Medical and Dental University, Tokyo 101,[*2] and Department of Physiology, Osaka University Medical School, Osaka 553,[*3] Japan*

Prostaglandin (PG) D_2 is identified as a natural constituent in brains of various mammals and its synthesis (6) and degradation (8) in brain have been investigated in detail. The central actions of PGE_2 and $PGF_{2\alpha}$ were widely examined and it has been demonstrated that several hypothalamic functions such as regulations of body temperature (3) and food intake are closely related to the production of PGE_2 or $PGF_{2\alpha}$ in the brain. However, the central actions of PGD_2, which is the major substance among prostaglandins in rat brain, have remained unclear. Recently, we found using the intracerebral microinjection technique that PGD_2 elicited several central actions. Although PGD_2 is an isomer of PGE_2, PGD_2 had specific actions on inducing hypothermia (9, 10) and bradycardia (10) which were quite different from the central actions of PGE_2. Another specific action of PGD_2 was the sleep-inducing effect (1, 2, 10, 11). In this paper, we demonstrate the results concerning the role of the prostaglandin on sleep regulation. Further, we discuss the possible role of endogenous PGD_2 in the regulation of physiological sleep.

I. INTRACEREBRAL INJECTION OF PGD₂

The experimental rat was placed on a small platform reported by
Prazma *et al.* (5). PGD_2 dissolved in saline was injected in 1 min into
the various regions of the rat brain through the guide cannula which had
been implanted 1 week before the experiment. The sleep stages were
determined by the polygraphic recordings of electroencephalogram
(EEG), electromyogram (EMG) and locomotion. Figure 1 shows the
sleep stage alteration after the microinjection of PGD_2 (2.5 nmol) to
the preoptic area of conscious rat. On the small platform, the control
rat (saline injected) remained awake for most of the experimental session of 200 min. On the other hand, PGD_2 induced slow wave sleep
(SWS) within several minutes after the injection. PGD_2 (0.2–1 nmol)
increased the amount of sleep in a dose-dependent manner. However,
paradoxical sleep (PS) was not observed under these conditions, probably due to the small platform on which the rats were placed and to
the short duration of the experiment. PGD_2 was not pyrogenic and it
slightly decreased the colonic temperature as is observed during physiological sleep. In an attempt to know the site of action of PGD_2, PGD_2
was injected stereotaxically to various regions in the brain. The site of
the injection was confirmed post-mortem. When PGD_2 was injected into
the preoptic area, the amount of sleep increased from the control level
of saline injection (Fig. 2). However, the injections into the cerebral
cortex, thalamus, locus coeruleus, and posterior hypothalamus had no

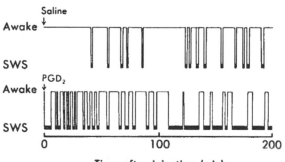

Fig. 1. Sleep stage alterations after the preoptic injections of saline or PGD₂ (2.5
nmol) dissolved in saline. Samples (3 μl) were injected in 1 min at time 0.

Ratio spent in different sleep stages (%)

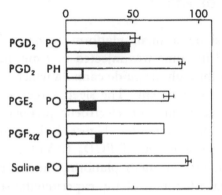

Fig. 2. The ratios (percent of recording time) spent in different sleep stages after the injection of prostaglandins. Prostaglandins (2.5 nmol) were injected into PO (preoptic area) and PH (posterior hypothalamus). The ratios were expressed by mean (%)±S.D. ($n=4$) for the experimental session of 3 hr. The ratio for PGF$_{2\alpha}$ was the mean (%) of two observations. ☐ awake; ▨ light SWS; ■ deep SWS.

effect. As shown in Figure 2, the injection into posterior hypothalamus, for example, did not increase the amount of sleep from the control level of saline injection. PGE$_2$ and PGF$_{2\alpha}$ slightly increased the amount of sleep but these actions were weaker than that of PGD$_2$. These results indicate that the preoptic area is a site of action of PGD$_2$ for inducing sleep. The preoptic area has been considered to be a center of sleep since Nauta reported in 1946 that section of the hypothalamus at a level just caudal to the preoptic area caused insomnia (*4*). Sterman and Clemente found that the electrical stimulation of the preoptic area caused SWS (*7*). The site of action of PGD$_2$ for inducing sleep coincided with the sleep center. In support of the results, Yamashita *et al.* (*13*) demonstrated histochemically high concentrations of PGD$_2$ receptors in this area.

II. SLEEP-INDUCING EFFECT OF PGD$_2$ IN THE DARK PERIOD

So far, we have presented the acute effect of PGD$_2$ on sleep under a high wake level of rats on the small platform. PGD$_2$ is an unstable compound which is degraded rapidly in the brain. To elucidate the sleep-inducing action of PGD$_2$ in more detail, we continuously infused a

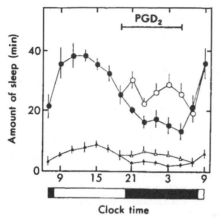

Fig. 3. Effects of nocturnal administration of PGD₂ on the amount of SWS and PS. Amounts of hourly SWS and PS per 2 hr (±1 hr of the indicated clock time) periods are plotted. Sleep patterns of control rats (●, SWS) and (▲, PS) under continuous saline infusion into the third ventricle ($n=10$). PGD₂ in sterile saline (0.36 nmol/200 μl) was infused into the third ventricle for 10 hr (19:00 to 5:00) at a rate of 600 fmol/min ($n=$ 4), and the amounts of SWS (○) and PS (△) were plotted. Throughout the infusion period, the rats were housed under the same conditions as the control rats without any restriction of the movement. Open and closed bars indicate light and dark periods, respectively. Values are means±S.E.M.

small amount of PGD₂ into the rat brain and monitored the daily sleep patterns. The circadian bioassay system was as follows (*11*, see also Chapter 17). After surgery the rat was housed in a cage placed in a soundproof room maintained at 25°C and 50% relative humidity on a 12 : 12 hr light-dark cycle. Food and water were available *ad libitum*. A slip ring allowed unrestrained movements of the rat. PGD₂ was infused continuously for 10 hr into the third ventricle of the brain. The sleep stages were determined for 96 hr on the basis of the polygraphic recordings of EEG, EMG, and locomotion. The sleep scores were computed and stored by a central processing unit. The behavior of the rat was monitored by a video-recording system. Under these conditions, circadian sleep patterns of rats were observed (closed circles in Fig. 3). The ordinate indicates the amounts of hourly SWS and PS, plotted for 2-hr epochs. Rats are night active animals as shown in Fig. 3. They sleep mainly in the light period. PGD₂ was infused at a rate of 600 fmol/ min during the active period at night. As indicated by open circles, PGD₂ increased the amount of both SWS and PS almost to the level of

TABLE I
Effects of Prostaglandins on Induction of Excess Sleep

Treatment	Rate (fmol/min)	n	SWS (% of control)	PS (% of control)
Saline				
(control)		20	100	100
PGD$_2$	6	3	96.5±2.3*	89.1±7.9*
	60	4	122.4±3.6**	110.3±11.9*
	600	8	133.3±3.9**	156.2±16.0***
	6,000	4	125.5±4.8**	178.3±12.1**
PGF$_{2\alpha}$	600	6	115.2±6.0***	100.4±16.3*
PGE$_2$	600	8	96.3±5.1*	104.3±11.7*

Prostaglandins were infused for 10 hr (19:00 to 5:00) and amounts of SWS and PS in the dark period (20:00 to 8:00) were determined. Results are mean±S.E.M. SWS, 228.1±5.8 min/12 hr; PS, 35.0±4.1 min/12 hr. Significance was calculated by t test.
*Not significant ($p > 0.05$), **$p < 0.001$, ***$p < 0.05$.

the light period. Sleep during infusion of PGD$_2$ was indistinguishable from physiological sleep as judged by EEG, EMG, and behavior. The effect of different doses of PGD$_2$ on sleep patterns at night was investigated (Table I). Although the infusion of PGD$_2$ at a rate of 6 fmol/min had no effect, as little as 60 fmol/min PGD$_2$ increased SWS significantly. The infusion of PGD$_2$ at doses higher than 600 fmol/min increased both SWS and PS. In addition, PGF$_{2\alpha}$ at 600 fmol/min increased SWS by 15% but the action was weaker than a corresponding dose of PGD$_2$. PGE$_2$ (600 fmol/min) enhanced neither SWS nor PS. Since the preoptic area contains about 4 pmol/g wet tissue (9), PGD$_2$ infused seems to be within a physiological range. On the basis of these results we propose that PGD$_2$ is an endogenous regulator of physiological sleep. If our hypothesis is correct, the amount of PGD$_2$ in the brain should change depending on the sleep/wake cycle.

III. THE CIRCADIAN RHYTHM OF PGD$_2$ SYNTHETASE AND PGD$_2$ DEHYDROGENASE IN RAT BRAIN

PGD$_2$ is synthesized from PGH$_2$ by PGD$_2$ synthetase (6) and degraded by PGD$_2$ dehydrogenase (8) in the brain from the fetal stage (12). We investigated the circadian changes of activities involved in PGD$_2$ metabolism. Brains were removed at the scheduled time and homogenized with a Polytron homogenizer. After centrifugation at $10,000 \times g$, the supernatant was obtained. This fraction contained both PGD$_2$ synthetase and PGD$_2$ dehydrogenase. Although the activity of PGD$_2$ dehydrogenase did not change over the light-dark cycle, PGD$_2$ synthetase

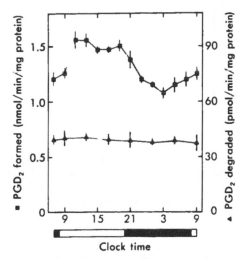

Fig. 4. Circadian changes of PGD$_2$ synthetase (■) and PGD$_2$ dehydrogenase (▲) in rat brain. The activity of PGD$_2$ synthetase was determined by tracing [^{14}C]-PGH$_2$ to [^{14}C]-PGD$_2$ (6). The activity of PGD$_2$ dehydrogenase was monitored spectrophotometrically by detecting 15-keto-PGD$_2$ at 415 nm (8).

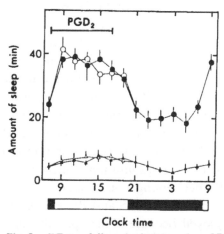

Fig. 5. Effects of diurnal administration of PGD$_2$ on the amount of SWS and PS. PGD$_2$ was administered into the third ventricle of rats for 10 hr (7:00–17:00) at a rate of 600 fmol/min. Control rats (●, ▲). PGD$_2$-infused rats (○, △). Details are described in Fig. 3. Values are means ± S.E.M.

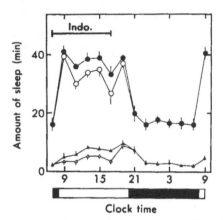

Fig. 6. Effect of diurnal administration of indomethacin. Indomethacin was infused continuously for 10 hr (7:00 to 17:00) through a cannula chronically implanted into the external jugular vein at a rate of 110 ng/min, and the amounts of hourly SWS (○) and PS (△) were plotted (*n*=6). Control SWS (●) and PS (▲). Values are means ± S.E.M.

exhibited circadian changes in the activity (Fig. 4). In the light period when the rat mainly sleeps, the activity of PGD_2 synthetase was higher than in the dark period. The circadian changes of PGD_2 synthetase activity showed a good correlation with the sleep/wake cycle. These results suggest that the endogenous production of PGD_2 may regulate the physiological sleep/wake rhythm depending on the light-dark cycle.

In order to examine this possibility, we exogenously administered PGD_2 (60 fmol–6 pmol/min) in the light period when endogenous PGD_2 is actively synthesized. As indicated in Table I, the nocturnal administration of PGD_2 increased both SWS and PS. However, the diurnal infusion of the same amount of PGD_2 failed to induce excess sleep (*2*) (Fig. 5), probably due to the high amount of the endogenous PGD_2. Inhibitors of prostaglandin biosynthesis were diurnally administered. Figure 6 shows the effect of a 10-hr intravenous infusion of indomethacin (110 ng/min) in the light period when endogenous PGD_2 is actively synthesized. Indomethacin decreased diurnal SWS and PS. Higher doses of indomethacin could not be employed because of the lipophilic nature of this compound. However, the amount of diurnal sleep was reduced to 50% of control by water soluble inhibitors of prostaglandin synthesis which decreased the prostaglandin level in rat brain. These results confirmed that the high sleep pressure in the light period is related to the formation of PGD_2 in the brain.

Several sleep substances have been reported. Among them, we found that muramyl dipeptide increased the level of PGD_2 in rat brain. This result indicated that the sleep-inducing effect of muramyl dipeptide might be mediated by the formation of endogenous PGD_2 in the brain.

Based on the all these results, we propose that PGD_2 in the brain could be an endogenous regulator of physiological sleep.

SUMMARY

We propose that PGD_2 in the brain may be an endogenous regulator of physiological sleep for the following reasons. 1) PGD_2 is a chemically defined natural constituent in the brain.   2) PGD_2 increases the amount of sleep in a dose-dependent manner. 3) A site of action of PGD_2 is the preoptic area. The preoptic area is recognized as a center of sleep. 4) Sleep induced by PGD_2 is indistinguishable from physiological sleep as judged by EEG, EMG, locomotor activity, and behavior of the rat. 5) As little as 60 fmol/min PGD_2 nocturnally administered causes excess sleep. The concentration of endogenous PGD_2 (4 pmol/g tissue) in the brain appears sufficient for inducing sleep. 6) PGD_2 is not pyrogenic and even slightly decreases body temperature as is observed in physiological sleep. 7) PGD_2 synthetase exhibits a circadian fluctuation in parallel to the sleep/wake cycle. In the light period when rats mainly sleep, PGD_2 is more actively synthesized than in the dark period. 8) In the light period when PGD_2 is actively synthesized in the brain, the administration of exogenous PGD_2 (60 fmol–6 pmol/min) fails to induce excess sleep. On the other hand, the administration of prostaglandin synthesis inhibitors decreases the amount of sleep.

REFERENCES

1 Inoué, S., Honda, K., Komoda, Y., Uchizono, K., Ueno, K., and Hayaishi, O. (1984). *Proc. Natl. Acad. Sci. U.S.* **81**, 6240–6244.

2 Inoué, S., Honda, K., Komoda, Y., Uchizono, K., Ueno, K., and Hayaishi, O. (1984). *Neurosci. Lett.* **49**, 207–211.

3 Milton, A.S. and Wendlandt, S. (1970). *J. Physiol.* **207**, 76–77.

4 Nauta, W.J.H. (1946). *J. Neurophysiol.* **9**, 285–316.

5 Prazma, J., Orr, J.L., and Kidwell, S.A. (1980). *Physiol. Behav.* **25**, 155–156.

6 Shimizu, T., Yamamoto, S., and Hayaishi, O. (1979). *J. Biol. Chem.* **254**, 5222–5228.

7 Sterman, M.B. and Clemente, C.D. (1962). *Exp. Neurol.* **6**, 103–117.

8 Tokumoto, H., Watanabe, K., Fukushima, D., Shimizu, T., and Hayaishi, O. (1982). *J. Biol. Chem.* **257**, 13576–13580.

9 Ueno, R., Narumiya, S., Ogorochi. T. Nakayama, T., Ishikawa, Y., and Hayaishi, O. (1982). *Proc. Natl. Acad. Sci. U.S.* **79**, 6093–6907.

10 Ueno, R., Ishikawa, Y., Nakayama, T., and Hayaishi, O. (1982). *Biochem. Biophys. Res. Commun.* **109**, 576–582.

11 Ueno, R., Honda, K., Inoué, S., and Hayaishi, O. (1983). *Proc. Natl. Acad. Sci. U.S.* **80**, 1735–1737.

12 Ueno, R., Osama, H., Urade, Y., and Hayaishi, O. (1985). *J. Neurochem.* **45**, 483–489.

13 Yamashita, A., Watanabe, Y., and Hayaishi, O. (1983). *Proc. Natl. Acad. Sci. U.S.* **80**, 6114–6118.

17

EFFECTS OF SLEEP-PROMOTING SUBSTANCES ON THE RAT CIRCADIAN SLEEP-WAKING CYCLES

KAZUKI HONDA,[*1] YASUO KOMODA,[*2] AND
SHOJIRO INOUÉ[*1]

*Divisions of Biocybernetics[*1] and Molecular Biology,[*2] Institute for Medical and
Dental Engineering, Tokyo Medical and Dental University, Tokyo 101, Japan*

A sleep-promoting factor was found in the brainstem of sleep-deprived rats (*14*). This factor, termed the sleep-promoting substance (SPS), largely modifies the circadian sleep-waking rhythm by inducing excess sleep, both slow wave sleep (SWS) and paradoxical sleep (PS), in the rat and the mouse (for an overview of SPS, see ref. *8*). Recent purification revealed that SPS contains at least four active components: SPS-A-1, SPS-A-2, SPS-B, and SPS-X (Fig. 1). SPS-A-1 was finally identified as uridine (*12*); the other components remain chemically unidentified. The present article deals with the somnogenic properties of uridine and SPS-B analyzed by our routine bioassay technique (*2*).

I. DONOR AND RECIPIENT ANIMALS FOR SPS

SPS was originally extracted from the brainstem of *ca.* 5,000 male rats purchased from local dealers. These donor rats were deprived of total sleep for 24 hr by being kept in our special sleep deprivation cages (*6*). After sleep deprivation, the rats were decapitated and their brainstems including the medulla oblongata, pons, mesencephalon, and hypothala-

203

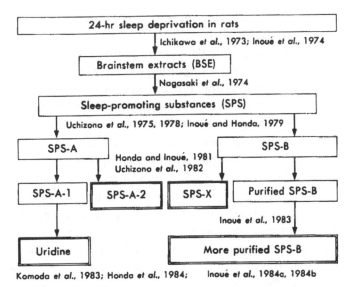

Fig. 1.   Steps of purification and isolation of SPS (from ref. *8*).

mus were immediately removed under ice-cooling. Autopsies were performed between 9:30 and 10:30. The tissues were homogenized with
and dialyzed against distilled water. The dialysate was lyophilized
and the residue yielded the brainstem extract (BSE). SPS was extracted
and purified from BSE (Fig. 1).

Using the routine sleep assay in rats (*2*) and mice (*15*), screenings
for active fractions have been performed since 1972. The BSE of 1,000
rats yielded 2.9 μg of uridine. The details of purification procedures are
described elsewhere (*12*).

For the evaluation of sleep-promoting activity of SPS, male rats
of the Sprague-Dawley strain, raised in our closed colony on a 12-hr
light and 12-hr dark schedule (light period: 8:00–20:00) under constant air-conditioned environment of $25\pm1°C$ and $60\pm6\%$ relative
humidity, were adopted as recipient animals.

Our rats exhibit a considerably stable circadian sleep-waking rhythmicity (*3*); hence these animals are especially suitable for a long-term
assay of sleep substances. Even under continuous intracerebroventricular
infusion of saline solution, rats regularly exhibited light-entrained circadian alternations in sleep amounts (*7*). Total time of SWS and PS
in the 12-hr environmental light period was respectively *ca.* 400 min

and *ca.* 70 min, while that in the alternating 12-hr dark period was respectively *ca.* 220 min and *ca.* 40 min.

At the age of 60–70 days, animals were implanted with electrodes for recording cortical electroencephalogram (EEG), neck electromyogram (EMG) and brain temperature. For infusion, a cannula was permanently inserted into the third ventricle. A cannular feedthrough slip ring fixed to the animal cage guaranteed free movements of the rats. A vibration transducer on the cage wall continuously detected the locomotor activity (*1*). Each cage was placed in a soundproof, electromagnetically shielded chamber under the same environmental conditions as described above. Physiological saline solution was continuously infused at a rate of 20 μl/hr throughout the observation period, except for a 10-hr period either at 19:00–5:00 (nocturnal infusion) or at 7:00–17:00 (diurnal infusion) during which rats were infused with a test solution. Uridine at 4 doses of 1, 10, 100, and 1,000 pmol, uracil at 3 doses of 1, 10, and 100 pmol, and SPS-B at 2 brainstem-equivalent units were dissolved in 200 μl saline and infused at the same rate of 20 μl/hr. Details of the experimental procedures are described in previous papers (*1, 2, 5*).

II. PROCESSING OF SLEEP DATA

The total assay system is schematically presented in Fig. 2.

The original EEG was filtered by frequency analyzers and delta waves (up to 4 Hz) were simultaneously recorded on the polygraphic sheets. Combining continuous records of locomotion, EEG, delta waves, and EMG, sleep/waking states were visually scored on a large-scale digitizer. The successive states of SWS, PS, and wakefulness were directly fed into and numerically processed by a computer-aided device system (*3*). The minimal scoring interval was 12 sec of recording time or 6 mm of polygraphic sheet. The criteria for visual scoring were described in a previous paper (*2*).

Wakefulness was characterized by a low-amplitude and high-frequency EEG, a high-amplitude EMG, the presence of locomotor activity, and the absence of delta waves. SWS was characterized by a high-amplitude and low-frequency EEG, a low-amplitude EMG, the absence of locomotor activity, and the presence of delta waves occupy-

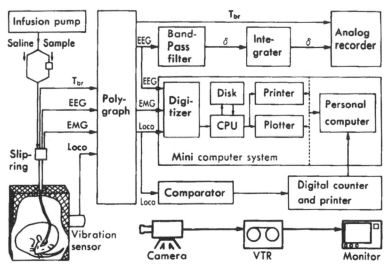

Fig. 2.   Total assay system (from ref. 8).

ing more than half of the minimal scoring interval. PS was characterized by a low-amplitude and high-frequency EEG, almost no trace of
EMG, the presence of intermittent slight locomotor activity and absence
of delta waves. PS occurred characteristically after an episode of SWS.
Any episode that lasted for less than 12 sec was added to the preceding
one and was not scored separately.

Sleep records on the baseline day were compared to those on successive days and statistically analyzed by Student's *t* test.

III.   SLEEP-PROMOTING EFFECTS OF URIDINE

1.   *Nocturnal Infusion*
As summarized in Fig. 3, a nocturnal infusion of 1–1,000 pmol uridine
exerted a dose-dependent effect in modulating sleep (5).

A 10-hr infusion of 1 pmol uridine exhibited no effect on the sleep-
waking patterns. Changes in total time of SWS and PS during the experimental night were less than 5% of that of the baseline night.
Average frequency and duration of episodes in wakefulness, SWS and
PS were not significantly different from those of the baseline in both the
light and the dark periods.

In contrast, a 10-hr infusion of 10 pmol uridine caused significant

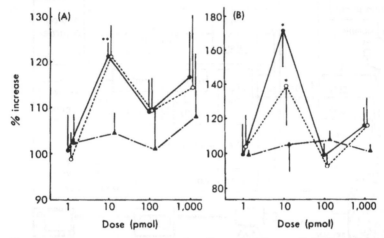

Fig. 3. Dose-response relations in the effects of uridine on SWS (A) and PS (B) in rats. Sleep parameters (mean±S.E.M.) in the experimental night were compared to those in the baseline night. ● total time; ○ frequency; ▲ duration. Single and double asterisk indicates that the difference from the baseline value is significant at $p < 0.05$ and $p < 0.01$, respectively (modified from ref. 5).

increases in both SWS and PS in the experimental night. The percent increase in total time of SWS and PS in the dark period was 21.0% (net increment: 45.5 min) and 68.1% (17.9 min), respectively, as compared to the baseline night. A similar increase was observed in the first recovery night: a 21.3% increase in SWS and a 55.9% increase in PS. Amounts of sleep returned to the baseline level on the second recovery day. The average frequency of episodes in wakefulness, SWS and PS also increased during the recovery nights. However, the average duration of both SWS and PS was almost the same as under baseline conditions. Hence, the average duration of wakefulness was shortened.

Time-course changes in hourly sleep amounts (Fig. 4) showed that the uridine infusion rapidly induced a slight but significant excess sleep in recipient rats. It was noted that after termination of the infusion excess sleep persisted, and that an almost similar sleep pattern resulted in the dark period of the following day. However, total amounts of sleep time and the other parameters during the light period remained stable and unchanged throughout the 4 consecutive days. Circadian and ultradian variations in brain temperature were little affected by

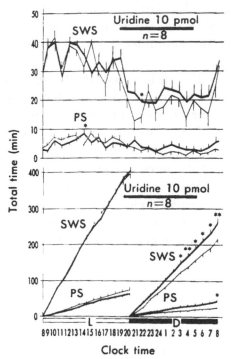

Fig. 4. Effects of a nocturnal infusion of 10 pmol uridine (indicated by a horizontal bar) on the time-course changes in SWS and PS in otherwise saline-infused rats. The top graph shows the hourly integrated sleep amounts, while the bottom graph illustrates the cumulative values in the environmental light (L) and dark (D) period. Thin and thick curves represent the baseline and the experimental day, respectively. Vertical lines on each hourly value indicate S.E.M. Single and double asterisk indicates that the value on the experimental day was significantly different from that of the baseline at $p < 0.05$ and $p < 0.01$, respectively (modified from ref. 10).

the uridine infusion. Neither a pyrogenic nor an antipyretic effect was detected (5).

A 10-hr infusion of 100 and 1,000 pmol uridine caused a slight and insignificant change in sleep amount in the dark period (Fig. 3). The dosage of 100 pmol exerted almost no effect on PS parameters, but elevated total time and frequency of SWS episodes by *ca.* 10%. Since the frequency of wakefulness episodes remained unchanged, their duration was shortened by 12%. The dosage of 1,000 pmol caused a severe disturbance in circadian locomotor activity rhythms in all recipient rats, 2 of which died within a few days. Consequently, this dosage was too high for intracerebroventricular administration and its effects can hardly be regarded as physiological. Radulovacki and Virus (*16*) also observed that 1–100 nmol of uridine administered intracerebroventricularly failed to cause excess sleep.

It is clearly demonstrated that the optimal dosage of uridine was 10 pmol in our nocturnal assay in freely moving rats. Lower and higher doses were ineffective. The SWS-promoting effect of uridine was rather mild and not so dramatic as that of other substances like delta-sleep-inducing peptide (DSIP), muramyl dipeptide (MDP), prostaglandin D_2 (PGD_2), and the other fractions of SPS tested by the same assay technique (*10*, *11*, see also Chapter 1). However, uridine was characterized by its long-lasting effect over 2 successive nights, and by a strong PS-promoting effect. Taking such effects into account, uridine might play a rather basic and nonspecific role in the regulation of sleep.

2. Diurnal Infusion

In contrast, a diurnal infusion of 1–1,000 pmol uridine resulted in little sleep-modulating effect. In the case of a 10 pmol infusion, as shown in Fig. 5, statistically significant fluctuations from the baseline level appeared in a few points of the hourly SWS and PS amounts, which were largely due to time-to-time variations but not directly related to the 10 pmol uridine infusion. The total time of SWS and PS in both the light and the dark period of the experimental day showed no significant difference from either the baseline day or the recovery days. The frequency and duration of SWS and PS episodes were little affected by uridine. No change was detected in brain temperature.

The ineffectiveness of the diurnal infusion of uridine is of particular interest from the viewpoint of the circadian mechanism of sleep regulation. DSIP and PGD_2 also exhibited the diurnal ineffectiveness (*9*). Hence, it is postulated that, during the light period, a physiological demand for sleep is so strong that the neurohumoral mechanism is maximally operative to fill the time available for rest with sleep. Consequently, in spite of the addition of an exogenously supplied sleep substance to the endogenously liberated humoral factors, no extra sleep can be induced (for discussion see also Chapter 1). In contrast, during the dark period, sleep is suppressed and replaced by wakefulness. The waking period can be easily transferred into the sleeping period by

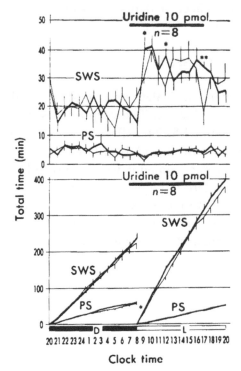

Fig. 5. Effects of a diurnal infusion of 10 pmol uridine. For explanation, see the legend of Fig. 4 (from ref. 9).

various factors. Consequently, a nocturnal infusion of sleep substances provoked excess sleep.

3. Nocturnal Infusion of Uracil

Uridine, a naturally occurring nucleoside, is known to be reversibly transformed to uracil, a base moiety of uridine, in the metabolic process. Hence the question arises as to whether uracil has also a sleep-modulating effect. A nocturnal infusion of 1–100 pmol uracil had little effect on the sleep-waking pattern (4). The total time of wakefulness, SWS, and PS in the light and dark period of day 2 after uracil administration did not differ significantly from either the baseline day or from the recovery days. The frequency and duration of wakefulness, SWS, and PS episodes were also little affected by the administration of uracil. This finding suggests that a sugar moiety of uridine may have an important role in the sleep-promoting effect or that some uridine metabolites other than uracil may be active.

IV. SLEEP-PROMOTING EFFECTS OF SPS-B

The nocturnal infusion of 2 brainstem-equivalent units of SPS-B resulted in a profound sleep-enhancing effect (Fig. 6). The sleep records showed a persistent increase in both SWS and PS. The effect appeared shortly after the initiation of the SPS-B infusion and lasted throughout the dark period. The hourly values in sleep amount were statistically different from the baseline at four different points for SWS and at one point for PS (Fig. 6, above). The cumulative values of SWS started to be significantly increased 5 hr after dark onset (Fig. 6, below). The total time of SWS in the dark period increased by 94.9 min (38.6%) as com-

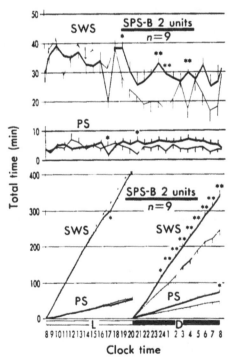

Fig. 6. Effects of a nocturnal infusion of 2 brainstem-equivalent units of more purified SPS-B. For explanation, see the legend of Fig. 4 (modified from ref. *10*).

pared to baseline (*10*, *11*). This was highly significant. The total time of PS also significantly increased, the increment being 23.8 min (49.9%). Hence, the net increase in total sleep time in the dark period was approximately 2 hr.

Similar to uridine, the increase in sleep was caused by the more frequent occurrence of SWS and PS episodes, while their duration was less affected. PS significantly decreased in the light period of the recovery day, as was also the case for the previous less purified fractions (*7*).

V. THE POSSIBLE ROLES OF SPS

It was found that uridine and SPS-B exhibited steady, long-lasting sleep-promoting effects. Both SWS and PS were affected. These SPS components seem to play a more or less similar basic role in triggering and maintaining sleep. The sleep enhancement was mainly due to the frequent occurrence of sleep episodes. Since natural sleep in rats is episodic and frequently interrupted by wakefulness, especially at the dark period, a prolongation of a single SWS and/or PS episode may not be physiological. In this respect, these "exogenously" supplied "endogenous" substances did not alter the natural sleep pattern.

Uridine was characterized by a slight but steady sleep-inducing and sleep-maintaining effect. PS was more affected than SWS. The optimal dosage was as small as 10 pmol. The infusion rate of 10 pmol/ 10 hr means that exogenously supplied uridine was liberated into the cerebrospinal fluid at a rate of 17 fmol/min. The cerebral action of such a low quantity of uridine has not yet been reported in the literature. Since uridine is rather widely distributed in the brain, it is postulated that an extremely slight elevation in cerebral uridine levels may be responsible for triggering sleep (*12*). In this connection, Inokuchi and Oomura (Chapter 18) found that electrophoretically applied uridine exerted a marked modulatory effect on the activity of rat hypothalamic preoptic neurons. Further studies will elucidate the neurohumoral mechanism associated with the sleep-promoting effect of uridine.

Some sleep substances are known to be related to the temperature regulation. MDP is pyrogenic (*10*, *13*). DSIP causes in rats both hypothermia and hyperthermia depending on the ambient temperature

(*18*). PGD$_2$ causes hypothermia in rats (*17*). However, uridine exerted no effect on brain temperature.

Among the materials tested by our assay technique, SPS-B was most potent in inducing excess sleep (*10, 11*, see also Chapter 1). Since SPS-B is still chromatographically impure, the content of the active material(s) in the test solution was uncertain. Nevertheless, a dosage of 2 brainstem-equivalent units of the donor rats indicates that SPS-B was effective at a considerably small dose. This factor was characterized by its steady, long-lasting effects with respect to both induction and maintenance of SWS and PS. If a chemically new compound is isolated from SPS-B, it might be a specific "sleep hormone" (*8*).

SUMMARY

SPS was extracted from the brainstem of 24-hr sleep-deprived rats. One of the effective fractions was finally identified as uridine. A 10-hr intracerebroventricular infusion of 10 pmol uridine at 19:00–5:00 caused a significant increase in sleep in the dark period (20:00–8:00) in otherwise saline-infused male rats ($n=8$). Increments of SWS and PS were 21.0% and 68.1%, respectively, as compared to the previous baseline night. This change was due to the increased frequency of both SWS and PS episodes with no change of their duration. A similar increase occurred in the first recovery night, but sleep amounts returned to the baseline levels on the second night. Brain temperature was not affected by uridine administration. A lower dose of uridine (1 pmol/10 hr) exerted no effect ($n=6$), while higher doses (100 and 1,000 pmol/10 hr, each $n=5$) caused a slight but insignificant increase in SWS and PS. In contrast, a 10-hr intracerebroventricular infusion of 10 pmol uridine at 7:00–17:00 caused no change in sleep ($n=8$). Uracil, a base moiety of uridine, similarly administered at doses of 1, 10, and 100 pmol (each $n=7$) at 19:00–5:00, exerted no sleep-modulating effect. Two brainstem-equivalents of SPS-B, a chemically unidentified fraction of SPS, which was infused at 19:00–5:00, caused dramatic increases in sleep in the dark period (38.6% in SWS and 49.9% in PS). Similar to the nocturnal infusion of 10 pmol uridine, the increase was due to the more frequent occurrence of SWS and PS episodes, while their duration was less affected. Thus, an extremely small dose of uridine and SPS-B induced excessive sleep when the intracerebroventricular administration started shortly before the onset of the dark period.

Acknowledgments

This study was supported in part by a Grant-in-Aid for Scientific Research to S.I. (No. 58480136) from the Ministry of Education, Science and Culture of Japan, and funds from Chugai Pharmaceutical Co., Ltd., Tokyo, the Suzuken Memorial Foundation, Nagoya, and the Naito Foundation, Tokyo.

REFERENCES

1 Honda, K., Ichikawa, H., and Inoué, S. (1974). *Rep. Inst. Med. Dent. Eng.* **8**, 149–152.
2 Honda, K. and Inoué, S. (1978). *Rep. Inst. Med. Dent. Eng.* **12**, 81–85.
3 Honda, K. and Inoué, S. (1981). *Rep. Inst. Med. Dent. Eng.* **15**, 115–123.
4 Honda, K., Komoda, Y., and Inoué, S. (1984). *Rep. Med. Dent. Eng.* **18**, 93–95.
5 Honda, K., Komoda, Y., Nishida, S., Nagasaki, H., Higashi, A., Uchizono, K., and Inoué, S. (1984). *Neurosci. Res.* **1**, 243–252.
6 Ichikawa, H., Honda, K., and Inoué, S. (1973). *Rep. Inst. Med. Dent. Eng.* **7**, 145–148.
7 Inoué, S., Honda, K., and Komoda, Y. (1983). In *Sleep 1982*, ed. Koella, W.P., pp. 112–114. Basel: Karger.
8 Inoué, S., Honda, K., and Komoda, Y. (1985). In *Sleep: Neurotransmitters and Neuromodulators*, ed. Wauquier, A., Gaillard, J.M., Monti, J.M., and Radulovacki, M., pp. 305–318. New York: Raven Press.
9 Inoué, S., Honda, K., Komoda, Y., Uchizobo, K., Ueno, R., and Hayaishi, O. (1984). *Neurosci. Lett.* **49**, 207–211.
10 Inoué, S., Honda, K. Komoda Y., Uchizono, K., Ueno, R., and Hayaishi, O. (1984). *Proc. Natl. Acad. Sci. U.S.* **81**, 6240–6244.
11 Inoué, S., Honda, K., Nishida, S., and Komoda, Y. (1983). *Sleep Res.* **12**, 81.
12 Komoda, Y., Ishikawa, M., Nagasaki, H., Iriki, M., Honda, K., Inoué, S., Higashi, A., and Uchizono, K. (1983). *Biomed. Res.* **4**, (Suppl.) 223–227.
13 Krueger, J.M., Pappenheimer, J.R., and Karnovsky, M.L. (1982). *Proc. Natl. Acad. Sci. U.S.* **79**, 6102–6106.
14 Nagasaki, H., Iriki, M., Inoué, S., and Uchizobo, K. (1974). *Proc. Japan. Acad.* **50**, 241–246.
15 Nagasaki, H., Kitahama, K., Valatx, J.-L., and Jouvet, M. (1980). *Brain Res.* **192**, 276–280.
16 Radulovacki, M. and Virus, R.M. (1985). In *Sleep: Neurotransmitters and Neuromodulators*, ed. Wauquier, A., Gaillard, J.M., Monti, J.M., and Radulovacki, M., pp. 221–227. New York: Raven Press.
17 Ueno, R., Narumiya, S., Ogorochi, T., Nakayama, T., Ishikawa, Y., and Hayaishi, O. (1982). *Proc. Natl. Acad. Sci. U.S.* **79**, 6093–6097.
18 Yehuda, S., Kastin, A.J., and Coy, D.H. (1980). *Pharmacol. Biochem. Behav.* **13**, 895–900.

18

EFFECTS OF PROSTAGLANDIN D₂ AND SLEEP-PROMOTING SUBSTANCE ON HYPOTHALAMIC NEURONAL ACTIVITY IN THE RAT

AKIRA INOKUCHI AND YUTAKA OOMURA

Department of Physiology, Faculty of Medicine, Kyushu University, Fukuoka 812, Japan

The preoptic area (POA) in the hypothalamus is deeply involved in central regulatory mechanisms of sleep. Stimulation of this area produced cortical electroencephalogram (EEG) synchronization and led to sleep (5). On the other hand, the suppression of sleep by POA lesion was observed in the rat (26) and cat (20). Recent work using a multi-unit recording technique revealed increased neuronal activity in the POA preceding the shift from waking to the slow wave sleep (SWS) state (8). Further, chemical stimulation experiments confirmed the importance of this region to sleep mechanism. Microinjection of noradrenaline (NA) into the POA (10) produces alertness, while microinjection of acetylcholine (ACh) elicits sleep (10). Microinjection of a sleep-inducing peptide into the basal forebrain at the level of the optic chiasm also induced excess SWS (7). The posterior hypothalamic area (PHA), too, is linked to the regulation of sleep. High frequency stimulation of this area induced a persistent EEG desynchronization in *cerveau isolé* cats (4) and lesion disturbed the function of waking in rats (26).

Since the classical hypnotoxin experiments of Legendre and Piéron (19), the existence of endogenous sleep-inducing factors has been proposed and a number of such substances have been reported (see ref. *14* for review). Systemic or intracerebroventricular administrations of prostaglandin D_2 (PGD_2) (*18, 36, 38*) or sleep-promoting substance (SPS) (*12, 24*) have been reported to induce excess sleep, and these substances are presumed to be sleep-inducing factors. In this paper, we report the effects of PGD_2 and SPS in the POA and PHA of the rat hypothalamus and interactions with several neurotransmitters.

I. PROSTAGLANDIN D_2

Prostaglandins (PGs) have significant roles in various tissues including the central nervous system (CNS). PGD_2 is a major PG in the brain (*1, 34*) and is highly concentrated in the pituitary, pineal body, hypothalamus and olfactory bulb (*25*). High concentrations of PGD synthetase, PGD_2 dehydrogenase and binding proteins specific for PGD_2 are present in the hypothalamus (*29–31, 35, 39*). Since these enzymes and binding proteins are rich in neural elements (*30, 39*), PGD_2 has been considered to be a neurotransmitter or neuromodulator and to have some physiological role in this region. In recent years, effects of PGD_2 in the CNS have been elucidated rapidly, and it has been reported that administration of PGD_2 into the CNS produced hypothermia (*37*), an increment of sleep (*36, 38*) and suppression of luteinizing hormone release (*16*).

The effects of PGs on CNS neurons have been also examined vigorously but the majority of such investigations were concerned with PGE and PGF. Direct effects of the latter on hypothalamic neuronal activity were mainly excitatory and the number of responsive neurons ranged from 9% to 83% (*15, 27, 33*). We examined the effects of PGD_2 on hypothalamic neurons using a microelectrophoretic technique, and almost half of the neurons tested in the POA and PHA responded. Excitatory and inhibitory effects were almost equally observed. In 7%, bidirectional responses also occurred that were similar to the response observed in PGF application on cerebellar Purkinje neurons (*32*).

Desensitization or tachyphylaxis was observed in 23% of POA neurons examined. In the cat, the effects of PGE and PGF series were frequently desensitized in brainstem neurons (*2*) but rarely in the hypothalamus (*15*). In the guinea pig, however, desensitization often occurred in hypothalamic neurons (*27*). In the rat, the neurons in the

cerebellar cortex, reticular formation, cerebral cortex, and hippocampus showed desensitization in only a few cases, and then after repetitive applications of PGE_1 (32). External application of PGD_2 to neuroblastoma × glioma hybrid cell produced long-lasting depolarization preceded by transient membrane hyperpolarization, and this effect was desensitized in almost all cases (11). It is suspected that desensitization is a general characteristic of PGs but the rate of occurrence depends on the species and the sites examined.

Antagonism of the NA response by PGE has been reported for cerebellar Purkinje neurons (32), but such an effect has not been observed on the hypothalamic neurons (15, 33). We observed modulatory effects of PGD_2 not only on NA response but also on dopamine (DA) and ACh responses. Modulatory effects of PGD_2 on NA response in the POA were very common (61%), and mainly the NA inhibitory response was attenuated, blocked or reversed during simultaneous PGD_2 application. An example of the latter is shown in the right part of Fig. 1. This neuron was initially inhibited by NA. PGD_2, itself, had no effect on spontaneous neuronal activity. During concurrent application of PGD_2 at $-50\sim-190$ nA, the NA inhibitory response changed to excitation (Fig. 1b and c). After termination of PGD_2, the NA response smoothly returned to inhibition (Fig. 1d). This modulation of the NA response seemed to be of two types: short- (20/23) and long-acting (3/23); a difference in sensitivity to PGD_2 or a different modulatory mechanism may account for the two types. Since both excitatory and inhibitory effects were blocked or reversed, it is suggested that both α- and β-adrenoceptive sites were affected by PGD_2. Recently, PGD_2 effects on sympathetic neurotransmission were reported in the anococcygeal muscle (3) and nictitating membrane (9). Facilitation of transmission occurred through pre- and post-junctional activation in the former, and inhibition occurred through pre-junctional inhibition in the latter. The fact that PGD_2 itself did not affect spontaneous neuronal activity in the experiments that examined PGD_2 modulatory effects suggests that it acts on the post-synaptic membrane. Other data showing that PGD_2 did not affect catecholamine release in rat cortex and striatal slices (28) support this idea. Moreover, neurotransmission in the POA following ventral noradrenergic bundle (VNB) stimulation was modulated by PGD_2 in 38% of neurons (6/16) tested; the left part of

Fig. 1 shows an example. Stimulation of the VNB resulted in inhibition (onset latency 9 msec, duration 82 msec) followed by excitation (duration 19 msec) (Fig. 1A). During PGD_2 application at -50 nA, inhibition was attenuated (onset latency 13 msec, duration 60 msec)

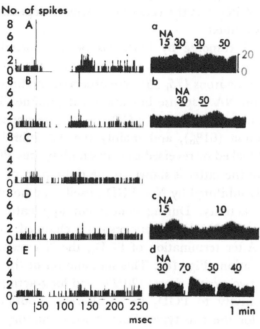

Fig. 1. Modulation of NA effect and neurotransmission in the POA by PGD_2. Left halves: poststimulus time histograms (PSTHs, 70 stimuli) of one POA neuron activity following ventral noradrenergic bundle stimulation (at arrow). All PSTHs consist of 512 × 0.5 ms bins. A: control, before PGD_2 application. Inhibition (onset latency 9 msec, duration 82 msec) followed by excitation (duration 19 msec). B: during PGD_2 at -50 nA. Note shift of excitation to the left. C: during PGD_2 at -130 nA. Note increased duration of excitation. D: during PGD_2 at -190 nA. Duration of inhibition, further reduced. Excitation unclear. E: 2 min after termination of PGD_2. Inhibition-excitation sequence reappeared. Dotted line, onset of excitation observed before PGD_2 application. Right halves: effects of NA just after each stimulation experiment. Ordinates: neuronal discharge, impulses/sec. Overbars: duration of electrophoretic application of indicated factor at current specified by arabic numeral. a: control, NA, no effect at 15 nA; inhibition followed by excitation at 30 nA; inhibition at 30 and 50 nA. b: during PGD_2 at -50 nA, NA, excitation at 30 and 50 nA. c: during PGD_2 at -190 nA, NA, excitation proceded by slight inhibition at 15 nA; excitation at 10 nA. Note remarkable excitation at low application current. d: recovery 2 min after termination of PGD_2, NA, inhibition at 30 nA; inhibition followed by excitation at 70 nA; inhibition at 50 and 40 nA.

and duration of excitation extended to 24 msec (Fig. 1B). During PGD₂ application at −130 nA, the duration of inhibition decreased to 46 msec and excitation extended to 108 msec (Fig. 1C). During PGD₂ application at −190 nA, the duration of inhibition was only 11 msec and that of excitation was not clear (Fig. 1D). Two min after termination of PGD₂, the inhibition (onset latency 11 msec, duration 77 msec) recovered and the onset of excitation was the same as that of A (Fig. 1E). PGD₂ application tended to reduce the duration of inhibition and to extend that of excitation. At the same time, the NA response on the same neuron was also modulated by PGD₂ (Fig. 1a to d). Since the inhibitory response of the lateral hypothalamic neuron following VNB stimulation and that of the POA neuron following medial forebrain bundle stimulation were blocked by noradrenergic antagonists (*13, 23*), modulation of noradrenergic neurotransmission in the POA by PGD₂ is highly suspected. The effects of DA in the POA were modulated in 63% of neurons (5/8) tested; mainly a marked attenuation of its excitatory or inhibitory effect was observed. The effects of ACh in the POA were also modulated by concurrent application of PGD₂ (38%), and the disappearance of excitation was dominant. The neurons, in which the ACh response was modulated by PGD₂, showed also modulation of the NA response by PGD₂. Thus, modulatory effects of PGD₂ of the NA response seem to be stronger than those of the ACh response. In contrast to the POA, modulations of NA, DA, and ACh responses were rarely seen in the PHA. The highly variable direct effect of PGD₂ on spontaneous activity of hypothalamic neurons and the potent modulation to catecholamine and ACh responses in the POA suggest that PGD₂ could act as a neuromodulator in this region.

As described in the introduction, the POA and PHA have significant roles in the regulation of the sleep-waking cycle. Since an increment of sleep as well as hypothermia were elicited by microinjection of PGD₂ into the POA and not by injection into the PHA (*36, 37*), the POA seems to be a site of action of PGD₂. PGD₂ had almost no modulatory effect of NA, DA, and ACh responses on the PHA neurons. Responses of NA and ACh in the POA were frequently modulated by PGD₂ in the direction of attenuation or reversal. But NA response modulation by PGD₂ occurred more frequently, and was more potent and long-lasting than the modulation of the ACh response. Therefore, the PGD_2 hypnogenic action might be partly due to a blockage or reversal of the NA awakening effect in the POA. This modulation also occurred in the PGD_2 infusion experiment into the third cerebral ventricle. In that experiment, we infused PGD_2 at doses of 50 to 150 nmol. When a certain peptide was injected into the cerebral ventricle, the brain parenchymal concentration was 1% of the injected dose 20 min after injection (6). If we assume the same rate, even though the substances were different, the concentration of PGD_2 in the infusion experiment might be comparable to that in the behavioral experiment in which PGD_2 induced excess sleep (36). The fact that PGD_2 significantly excited neurons that were excited by ACh, might also contribute to the increment of sleep produced by PGD_2, judging from the ACh hypnogenic effect observed in microinjection experiments (10). It is concluded that PGD_2 is a neuromodulator which probably modulates noradrenergic neurotransmission in the hypothalamus. Through this mechanism, PGD_2 could be important for the sleep-waking regulation in the POA.

II. SLEEP-PROMOTING SUBSTANCE

SPS has been extracted from the brainstem of sleep-deprived rats (22) and its infusion into the rat cerebral ventricle (12) or the mouse peritoneum (24) increased the duration of sleep. Recently, the effective component of SPS was identified as uridine (UR) by Komoda et al. (17). The effect of SPS has been examined on the crayfish abdominal ganglion, where it inhibits spontaneous discharges (23). We first observed the effects of SPS on any neuron in the CNS, and almost one fourth of the neurons tested in the POA and PHA showed either excitatory or inhibitory responses. The effect of UR was comparable to that of SPS in 87% of the tested cases. Therefore, our electrophysiological data support UR as the effective component of SPS. The effects of SPS in the POA were closely related to those of glutamate (Glu), NA, and DA. Moreover, the effects of these neurotransmitters, except DA, were sometimes (24–32%) modulated by concurrent application of SPS in the POA. A modulatory effect of SPS of the NA response is shown in Fig. 2. This neuron responded to NA with inhibition and to DA and Glu with excitation. The NA response of this neuron changed to excita-

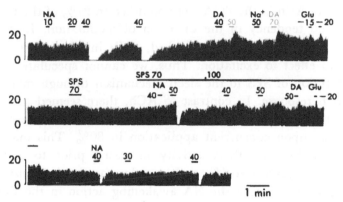

Fig. 2. SPS modulation of NA effect on POA neuronal activity. Continuous from upper to lower trace. Upper: inhibition by NA at 40 nA, excitation by DA at 50 and 70 nA and Glu at −15 and −20 nA, NA at 10 and 20 nA, DA at 40 nA and Na⁺ at 50 nA showed no effect. Middle: SPS at 70 nA showed no effect. During SPS application (70 nA), NA at 40 nA exerted no effect, but there was inhibition preceded by slight excitation at 50 nA. After increased concurrent SPS application at 100 nA, NA (40, 50 nA) effect was modulated from inhibition to excitation. Effects of DA (50 nA) and Glu (−20 nA) were not modulated. Lower: after termination of SPS application, inhibitory effect of NA (40 nA) recovered but was still attenuated. NA at 30 nA showed no effect (15).

tion during concurrent SPS application. At this time, DA and Glu were still excitatory. Several min after termination of the SPS application, the NA inhibitory response recovered partially but was still weak. Responses to 5-hydroxytryptamine (5-HT) changed in 25% of the tested cases. These findings suggest the possibility that UR, which is an effective component of SPS, may be a neuromodulator. Concentration, distribution, localization, and metabolism of UR in the brain is still an open question, on which we cannot comment at present. Modulatory effects of SPS were also seen in the PHA, but the occurrence was less frequent (modulation rates in the POA vs. those in the PHA: Glu, 32% vs. 21%; NA, 24% vs. 17%; ACh, 21% vs. 8%). The difference in modulation rate may depend on differences in neuronal sensitivity to SPS in the examined regions. It required several applications to exert the modulatory effects after the start of SPS application and to recover the original response levels after termination of application (more than 10 min). It seems probable that the SPS modulatory effects are mediated by metabolic processes rather than by participation in certain receptor mechanisms.

The effects of ACh in the POA were modified in 21% of the trials, but there was no tendency in the direction of modulation. In some cases, the excitatory responses were blocked, while in others there was change from no effect to excitation. Thus, we cannot speculate on a possible contribution of SPS to the sleep mechanism through modulation of the ACh response. In contrast to ACh, there seemed to be a tendency toward modulation of NA responses by SPS. SPS reversed the NA responses upon concurrent application in 80%. This may be related to the finding that POA activity increased prior to a shift from waking to the SWS state (8), and to a decrease in the conscious level by blocking or reversing the NA awakening action in the POA. Long-lasting modulation of neurotransmitter responses might reflect long-lasting sleep-promoting effects observed in behavioral experiments (14). We have, at present, no idea about possible contributions of the modulation of Glu and 5-HT responses in the hypothalamus during sleep.

It is pertinent to conclude that uridine modulates neurotransmission of Glu, NA, and possibly ACh and 5-HT in the hypothalamus, and this modulation may occur via metabolic processes. Although we did not examine the effects of SPS in other sites involved in sleep mechanisms (i.e., the dorsal raphe, locus coeruleus, and pontine reticular formation), modulation of NA responses in the POA might partly contribute to the regulation of the sleep-wakefulness mechanism.

SUMMARY

The effects of electrophoretically applied PGD_2, SPS, and UR on neuronal activity of the POA and PHA, and their interactions with several neurotransmitters were examined in the rat. In the POA, 20% of the tested neurons were excited, 26% were inhibited and 6% showed a bidirectional response to PGD_2. The excitatory effect of ACh was sometimes attenuated, blocked or reversed by concurrent PGD_2 application. The inhibitory response of NA frequently changed to no effect or to excitation during simultaneous PGD_2 application. Furthermore, a similar modulation was observed after PGD_2 infusion into the third cerebral ventricle. Neurotransmission in the POA following ventral noradrenergic bundle stimulation was modified in 38% by PGD_2. The direct effects of

PGD_2 in the PHA were similar to those in the POA, but modulations of ACh and NA responses by PGD_2 were rarely seen in the PHA.

SPS application excited POA neurons in 14% and inhibited in 13%. The effect of UR was comparable to that of SPS. The excitatory effect of Glu was blocked or attenuated in 30% by concurrent application of SPS, and this modulatory effect was long-lasting. The NA inhibitory effect was attenuated, blocked or reversed in 28% by SPS. The effects of ACh were modified in $21°_0$ by SPS. The direct effect of SPS in the PHA was similar to that in the POA. In these neurons, modulations of Glu and NA responses were observed during concurrent SPS application, but less frequently than in the POA. A possible contribution of PGD_2 and SPS in the hypothalamus to the sleep mechanism is discussed.

Acknowledgments

We thank Prof. A. Simpson for help in preparing this manuscript, Prof. O. Hayaishi for providing us with the PGD_2, and Drs. Y. Komoda and H. Nagasaki with the SPS and UR used in this study. Also, thanks are due to the Ono Pharmaceutical Co. Ltd. for the kind donation of PGD_2. This work was partly supported by Grants-in-Aid for Scientific Research (Y.O.) No. 5744085, 58370006, and 58870118 from the Ministry of Education, Science and Culture of Japan.

REFERENCES

1 Abdel-Halim, M.S., Hamberg, M., Sjöquest, B., and Änggard, E. (1977). *Prostaglandins* **14**, 633–643.
2 Avanzino, G.L., Bradley, P.B., and Wolstencroft, J.H. (1966). *Br. J. Pharmacol. Chemother.* **27**, 157–163.
3 Bedwani, J.R. and Hill, S.E. (1980). *Br. J. Pharmacol.* **69**, 609–614.
4 Belardetti, F., Borgia, R., and Mancia, M. (1977). *Electroenceph. Clin. Neurophysiol.* **42**, 213–225.
5 Clemente, C.D. and Sterman, M.B. (1967). In *Progress in Brain Research*, ed. Adey, W.R. and Tokizane, T., vol. 27, pp. 34–47. Amsterdam: Elsevier.
6 De Wildt, D., Verhoef, J., and Witter, A. (1982). *J. Neurochem.* **38**, 67–74.
7 García-Arrarás, J.E. and Pappenheimer, J.R. (1983). *J. Neurophysiol.* **49**, 528–533.
8 Hanada, Y. and Kawamura, H. (1984). *J. Physiol. Soc. Japan* **46**, 416.
9 Hemker, D.P. and Aiken, J.W. (1980). *Prostaglandins* **20**, 321–332.
10 Hernández-Peón, R., Chávez-Ibarra, G., Morgane, P.J., and Timo-Iaria, C. (1963). *Exp. Neurol.* **8**, 93–111.

11 Higashida, H., Nakagawa, Y., and Miki, N. (1984). *Brain Res.* **295**, 113–119.

12 Honda, K. and Inoué, S. (1978). *Rep. Inst. Med. Dent. Eng.* **12**, 81–85.

13 Hori, T., Kiyohara, T., Nakashima, T., and Shibata, M. (1982). *Brain Res. Bull.* **8**, 667–675.

14 Inoué, S., Uchizono, K., and Nagasaki, H. (1982). *Trends Neurosci.* **5**, 218–220.

15 Jell, R.M. and Sweatman, P. (1977). *Can. J. Physiol. Pharmacol.* **55**, 560–567.

16 Kinoshita, F., Nakai, Y., Katakami, H., Imura, H., Shimizu, T., and Hayaishi, O. (1982). *Endocrinology* **110**, 2207–2209.

17 Komoda, Y., Ishikawa, M., Nagasaki, H., Iriki, M., Honda, K., Inoué, S., Higashi, A., and Uchizono, K. (1983). *Biomed. Res.* **4** (Suppl)., 223–227.

18 Laychock, S.G., Johnson, D.N., and Harris, L.S. (1980). *Pharmacol. Biochem. Behav.* **12**, 747–754.

19 Legendre, R. and Piéron, H. (1910). *C.R. Soc. Biol. (Paris)* **68**, 1077–1079.

20 McGinty, D.J. and Sterman, M.B. (1968). *Science* **160**, 1253–1255.

21 Miyahara, S. and Oomura, Y. (1982). *Brain Res.* **234**, 459–463.

22 Nagasaki, H., Iriki, M., Inoué, S., and Uchizono, K. (1974). *Proc. Japan Acad.* **50**, 241–246.

23 Nagasaki, H., Iriki, M., and Uchizono, K. (1976). *Brain Res.* **109**, 202–205.

24 Nagasaki, H., Kitahama, K., Valatx, J.-L., and Jouvet, M. (1980). *Brain Res.* **192**, 276–280.

25 Narumiya, S., Ogorochi, T., Nakao, K., and Hayaishi, O. (1982). *Life Sci.* **31**, 2093–2103.

26 Nauta, W.J.H. (1946). *J. Neurophysiol.* **9**, 285–316.

27 Poulain, P. and Carette, B. (1974). *Brain Res.* **79**, 311–314.

28 Reimann, W., Steinhauer, H.B., Hedler, L., Starke, K., and Hertting, G. (1981). *Eur. J. Pharmacol.* **69**, 421–427.

29 Shimizu, T., Yamamoto, S., and Hayaishi, O. (1979). *J. Biol. Chem.* **254**, 5222–5228.

30 Shimizu, T., Mizuno, N., Amano, T., and Hayaishi, O. (1979). *Proc. Natl. Acad. Sci. U.S.* **76**, 6231–6234.

31 Shimizu, T., Yamashita, A., and Hayaishi, O. (1982). *J. Biol. Chem.* **257**, 13570–13575.

32 Siggins, G., Hoffer, B., and Bloom, F. (1971). *Ann. N.Y. Acad. Sci.* **180**, 303–323.

33 Stit, J.T. and Hardy, J.D. (1975). *Am. J. Physiol.* **229**, 240–245.

34 Sun, F.F., Champman, J.P., and McGuire, J.C. (1977). *Prostaglandins* **14**, 1055–1074.

35 Tokumoto, H., Watanabe, K., Fukushima, D., Shimizu, T., and Hayaishi, O. (1982). *J. Biol. Chem.* **257**, 13576–13580.

36 Ueno, R., Ishikawa, Y., Nakayama, T., and Hayaishi, O. (1982). *Biochem. Biophys. Res. Commun.* **109**, 576–582.

37 Ueno, R., Narumiya, S., Ogorochi, T., Nakayama, T., Ishikawa, Y., and Hayaishi, O. (1982). *Proc. Natl. Acad. Sci. U.S.* **79**, 6093–6097.

38 Ueno, R., Honda, K., Inoué, S., and Hayaishi, O. (1983). *Proc. Natl. Acad. Sci. U.S.* **80**, 1735–1737.

39 Yamashita, A., Watanabe, Y., and Hayaishi, O. (1983). *Proc. Natl. Acad. Sci. U.S.* **80**, 6114–6118.

PS-FACTORS AND
NEUROTRANSMITTERS

19

PS-INDUCING FACTORS AND THE NORADRENERGIC SYSTEM

JOËLLE ADRIEN AND CHRISTINE DUGOVIC

INSERM Pitié-Salpêtrière, 75013 Paris, France

At the beginning of the century, it was proposed that endogenous hypnogenic substances are present in the cerebrospinal fluid (CSF) of mammals (*18*). Since that time, other observations have supported this hypothesis: in particular the sleep rebound which follows sleep deprivation suggests that sleep-inducing factors accumulate in the central nervous system (CNS) during the deprivation period. Moreover, this rebound effect is also observed specifically for the state of paradoxical sleep (PS) after selective PS deprivation (*3, 24*). Other evidence for the presence of hypnogenic substances in brain and/or CSF was provided by several groups who showed: i) that sleep substances were accumulating in the venous blood of rabbits submitted to stimulation of the intralaminar thalamic nuclei which provokes cortical delta waves (*12*); ii) that hypnogenic factors were present in the CSF of sleep-deprived goats (*17*) and cats (*21*), and iii) that they were also found in perfusates of the midbrain reticular formation of sleeping cats (*4*) and in brainstem extracts of sleep-deprived rats (*15*).

More precisely, it was shown that the transfer of CSF from a sleep deprived donor to recipient rabbits or rats increased slow wave sleep

227

(SWS) in the latter (*2, 5, 16*). Similarly, in cats pre-treated with para-chlorophenylalanine (PCPA), SWS and PS were induced by intra-cerebroventricular (i.c.v.) infusion of CSF from PS-deprived donor cats (*21*). This last experiment, together with others performed by the same group (*8*), suggests that some sleep factor(s) are synthesized and/or liberated in the CSF of cats under the control of the serotonergic system.

Therefore, the question was raised here whether the well-known role of noradrenaline (NA) in the regulation of sleep (*6*) could also involve some sleep factors. In fact, there is general agreement that the normal occurrence of PS requires a certain functional state of the NA system, at both the pre- and the post-synaptic levels (*6, 7, 11*). If this NA regulation operates through controlling the synthesis and/or liberation of sleep factors in the CSF, then these factors should be able to induce sleep in animals rendered insomniac by a blockade of the NA system.

I. THE CHOICE OF A MODEL

Considering that each sleep state has its own characteristics and regulations, it is probable that SWS and PS are controlled by different hypnogenic substances (*8*). With this in mind, it was decided in the present experiments on CSF transfers to work with donor and recipient animals in which the experimental variable was the state of PS.

In fact, as far as the donor was concerned, it was taken into account that PS is the only sleep state which can be selectively suppressed, and that an instrumental PS deprivation is followed by a well-known PS rebound (*24*), which could correspond to the accumulation of PS-inducing factors in the brain. It was therefore hypothesized that the CSF of a PS-deprived donor rat contained substances which should induce PS in a recipient rat. To observe such an effect more clearly, we selected as the recipient rat an animal in which PS had been specifically decreased by pretreatment with a NA blocking agent.

For several reasons, primarily the propranolol (PRO)-induced insomnia model (*9, 10*) was used in the recipients:
The sleep deficit induced by PRO is essentially a PS decrease with no modification of SWS amounts.

The PS deficit is specifically due to the central blockade of the β_1-adrenoceptors (*11*).

The PS deficit can be antagonized by β-agonists such as isoproterenol and prenalterol. However, and contrary to what is observed with PCPA (*20*), the PRO effects on PS seem not to be reversed by peptides (this study) or by other compounds acting on monoaminergic receptors (β_2- or 5-hydroxytryptamine (5-HT)-agonists) (*11*).

In summary, after PRO treatment, the PS amounts are directly related to the blockade of the adrenoceptors. If the CSF transferred is able to counteract this effect of PRO, it schematically can be due either to its β_1-adrenergic agonist properties or to its action downstream from NA in the "sequence of biochemical reactions leading to the final cause of PS" (*8*). The latter alternative represents the hypothesis which was tested in the present study.

II. CSF FROM PS-DEPRIVED RATS INDUCES PS IN RECIPIENTS

All animals were implanted with the usual electrodes for sleep monitoring. In addition, cannulas were stereotaxically positioned in the 4th and/or the lateral ventricle in order to subsequently puncture or infuse CSF. The rats were housed in individual recording cages and lived under a 12–12 hr light-dark cycle (light on at 7:00), with food and water *ad libitum*.

The donor rats underwent a selective PS deprivation during 1 to 4 days, using the flower pot technique (*14*) for most of the time, and manual awakenings during 2 to 6 hr per day. This mixed deprivation procedure allowed the animals to have minimal stress, as estimated by their reactions while being handled, and by their body weight which was maintained or even increased during the deprivation period. This group of rats was deprived of 93% of PS, and of only 6% of SWS.

A group of "stress control" rats underwent the same procedure with a larger flower pot where they could sleep almost normally. They were not deprived of PS nor of SWS.

At the end of the deprivation period, 20 to 40 μl of CSF were punctured from the cerebroventricular space through the cannulas at a rate of about 2 μl/min, using a Hamilton syringe. Then the donor rat

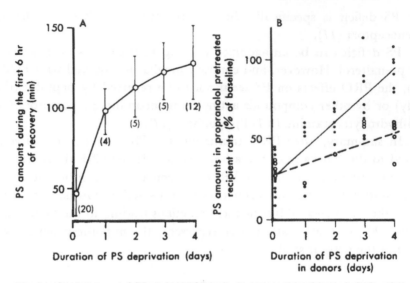

Fig. 1. A: Amounts of PS, expressed in min (mean and standard deviation for a total of 14 animals, number of tests in brackets), during the first 6 hr of the recovery period after PS deprivation of different durations. The typical PS rebounds observed suggest a progressive accumulation of some PS factor. B: PS restoration in PRO-pretreated recipient rats following transfer of 20 µl of fresh (●) and of frozen (−20°C) (○) CSF from PS-deprived donors. PS amounts are calculated for the 6-hr period following the transfer, and are expressed as percent of baseline recordings where the animals received dose 0 of PRO and 20 µl of artificial CSF (the average PS amount in these baseline conditions was 39 min per 6 hr. The circles represent individual tests, and the lines are the regressions for each group (fresh CSF: $r=0.8$, $p<0.001$, frozen CSF: $r=0.74$, $p<0.01$). Note that there is a "dose"-dependent effect for fresh CSF (slope=0.66), whereas freezing markedly decreased this action (slope=0.26).

was placed back into its standard recording cage and was allowed to sleep.

The PS-deprived groups ($n=24$) exhibited a typical PS rebound during the first 6 to 8 hr-period of recovery from deprivation (Figs. 1 and 2), whereas SWS was not changed (not shown). This rebound increased with the duration of the preceding deprivation, suggesting that the level of hypothetical sleep-inducing substances in the brain increased in a like manner. Therefore, the effects obtained in the recipient rats were analyzed with dose-response graphs.

In the stress-control group ($n=3$), there was no significant PS rebound (PS amounts per 6 hr=102% of baseline, and SWS=93%),

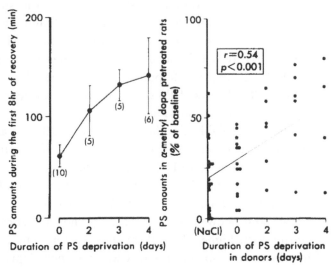

Fig. 2. A: Amounts of PS, expressed in min (mean and standard deviation for 10 animals, number of tests in brackets), during the first 8 hr of the recovery period after PS deprivation of different durations. Same comments as in Fig. 1. B: PS restoration in α-methyldopa (αMD)-pretreated recipient rats following transfer of 20 μl CSF from PS-deprived donors. PS amounts are calculated for the 8 hr-period following the transfer, and are expressed as percent of baseline recordings where the animals received dose 0 of αMD and 20 μl of artificial CSF (the average PS amount in these baseline conditions was of 63 min per 8 hr). Dots represent individual tests, and the line the regression line. Note that there is a significant "dose"-dependent effect of the CSF transfer.

suggesting that no sleep inducing factors had accumulated in these animals.

The *recipient rats* underwent two different types of pharmacological pre-treatment which both impaired the NA system and led during the 6–10 hr post-injection to a significant PS decrease and no change in SWS.
1) The PRO-induced PS insomnia was obtained by using an intraperitoneal injection of 10 mg/kg of D,L-PRO at 11:30 (*9*). At 12:00, the rats received an infusion into the 4th ventricle of 20 μl of CSF from a donor which had been deprived of PS during 1 to 4 days (corresponding respectively to doses 0 to 4).

For this procedure, the animals were sitting in their cage, while the probe of a Hamilton syringe was positioned inside the cannula. Twenty microliters of CSF was infused slowly into the ventricle over a few minutes.

With CSF from non-deprived donors (dose 0) or from the stress control group, the recipient rats exhibited during 6 hr a PS deficit which was the same as that observed after PRO treatment alone (9). However, when the recipients received CSF at doses 1 to 4, they exhibited increasing amounts of PS during the 6 post-infusion hours (Fig. 1), with no change in SWS. After 4 days of PS deprivation, the amounts of PS in the recipient rats reached normal values.

This restorative effect was highly significant ($r=0.8$, $p<0.001$) and was due to an increasing number of PS phases with no modification of their mean duration. On the other hand, the kinetics of this effect consisted of a maximum action of the CSF transfer during the 2 to 6 hr-period after the infusion. The latency of the first PS episode after treatment was very variable and showed a trend, though not statistically significant, to be shorter with higher "doses" of sleep factor. The long latencies of 4 to 5 hr which were observed with CSF from non-deprived donors were never found in experiments using CSF from 4-day PS-deprived rats where the longest latencies were 1 to 2 hr.

2)   *The α-methyldopa (αMD)-induced insomnia* was obtained by using 50 mg/kg of DL-αMD injected intraperitoneally at 9:30. This treatment provoked during 9 to 10 hr an 80% PS decrease with no important modifications of the total SWS amounts (SWS2 decreased by 20% whereas SWS1 increased by 30%). However, a state of sedation was observed during the first post-injection hour, and for this reason the i.c.v. infusion of CSF was performed only 90 min later, in order to deal with actual PS insomnia rather than with the immediate sedation following αMD.

The αMD-pretreated rats were infused in the 4th ventricle with 20 μl of CSF from donors deprived of PS during 2 to 4 days, according to the previously described protocol. The recipients were recorded for 8 to 20 hr thereafter, and the effects obtained on sleep states were analyzed in dose-response terms.

With CSF from non-deprived donors, the recipient rats exhibited the same PS deficit as with αMD alone. But when they were infused with CSF from PS-deprived donors, a progressive increase of PS amounts was observed, which depended on the duration of the deprivation in the donor ($r=0.54$, $p<0.001$). For the doses 3 and 4 of CSF, PS in the recipient rats attained about 60% of baseline (Fig. 2). This restoration

TABLE I
Effects of CSF Transfer on SWS in αMD Pre-treated Rats

		Duration of the deprivation (days)			
		0	2	3	4
Total SWS	0–8 hr	115±7	111±16	104±16	96±8
	0–4	115±12	110±27	105±22	100±15
	4–8	116±12	112±9	103±19	91±11
SWS 1	0–8 hr	150±37**	147±47*	132±22*	111±31
	0–4	180±60**	172±66*	169±43**	137±27*
	4–8	124±32*	125±43	94±11	80±41
SWS 2	0–8 hr	99±20	99±39	90±24	90±8
	0–4	90±29	93±48	77±29	86±14
	4–8	110±21	107±32	107±30	96±6

The amounts of SWS are expressed as percent of the baseline conditions for the 0–8-hr period after the transfer (SWS1 =98 min, SWS2 =201 min per 8 hr), as well as for each of the two 4-hr periods. Note that the amounts of total SWS and of SWS2 were not significantly modified in any condition. On the contrary, SWS1 was increased following αMD, especially during the first 4 hr post-treatment, and decreased progressively with larger "doses" of CSF factors. Student t test in comparison to baseline values: ** $p<0.01$, * $p<0.05$.

was achieved by an increase in the number of PS phases with no modification of their mean duration. On the other hand, the main effect of the transfer was observed during the 4 to 8 hr-period after the infusion, as indicated by the slope of the regression lines (0.27 as compared to 0.13 during the 0–4-hr period post-infusion) (not shown). With regard to SWS, there was no change in its total amount nor in SWS2, except for a tendency for SWS1 to decrease to its control value with increasing durations of deprivation in the donor (Table I).

3) *In summary*, the transfer of CSF from a PS-deprived donor to a recipient rat induced a restoration of PS which had been blocked by a pretreatment with pharmacological agents impairing the NA transmission. The efficiency of the CSF transferred was directly linked to the duration of the deprivation in the donor. Moreover, this action seemed to be specific for PS deprivation *per se* because the CSF from stress-control rats did not induce sleep in the recipients.

There are two alternative explanations for the PS-inducing effects obtained in the present experiments. The first one proposes that the mechanisms responsible for this phenomenon are strictly concerned with NA transmission by a direct stimulation of the adrenoceptors. According to this interpretation, the restoration of PS in recipient rats would be due only to the accumulation of NA in the CSF of donors, which acts as an agonist to stimulate the receptor sites. However, this hypothesis can be discarded, as it was demonstrated that NA levels in the brain are not modified by PS deprivation (*23*). Even when the NA turnover was increased (*19, 22*), such an effect was limited to the synaptic sites and did not involve the CSF.

The second alternate explanation is that sleep-inducing factors of non-noradrenergic nature accumulate in the CSF of PS-deprived rats, and that such substances by-pass the noradrenergic step in the central chain of biochemical events leading to PS. In this case, it should be noted that these factors are quite potent since the "dose" which was transferred represents only approximately one tenth of the total amount contained in the CSF of the donor. Moreover, if these factors accumulate preferentially in brain tissue rather than in the CSF, the part which was transferred would in fact represent only a very small fraction of the amount which was present in the donor animal.

This could explain why in previous studies the transfer of CSF from sleep-deprived to normally-sleeping animals has failed to cause a major sleep induction, even though the donor animal exhibited an obvious sleep rebound. While the total amount of sleep-inducing factors can affect the donor, only a small part is available for the recipient. According to this argument one would expect that if a non-pretreated recipient rat were given all the available sleep factors contained in the CNS of a PS-deprived donor, it should also demonstrate a PS rebound. This will be tested in our laboratory by attempting to concentrate the CSF.

III. TOWARD CHARACTERIZATION OF THE SLEEP-INDUCING SUBSTANCES

In our model, we assume that some sleep factors contained in the CSF are responsible for inducing PS under certain conditions of pharmacological pretreatment. The first question to ask is whether these factors correspond to those which have been already studied by other groups.

It was therefore attempted to induce PS in PRO-treated rats by using delta-sleep-inducing peptide (DSIP) (*13*). Presently, these experiments are in progress, and the preliminary results seem to indicate that DSIP (i.v. 25 μg/kg, or s.c. 100 μg/kg) does not restore PS after PRO-treatment, after a single or after repeated injections.

To further characterize the PS-inducing factors in the present model, their stability was assessed. We stored the punctured CSF for 1 to 15 days at a temperature of −20°C. Then it was brought to room temperature and immediately infused i.c.v. to recipient rats pretreated with PRO. The data obtained indicated that there was much less sleep-inducing potency after freezing (slope=0.26 as compared to 0.66 with fresh CSF, $r=0.74$, $p<0.01$) (Fig. 1). However, more recent data showed that storage at −80°C did not induce a loss of sleep-inducing properties. This latter point is very important because stability is a prerequisite for a further characterization of these factors.

SUMMARY

It was demonstrated that CSF from rats selectively deprived of PS was able to induce PS in rats that had been rendered insomniac by a pretreatment with PRO or α-MD. This phenomenon, together with the PS rebound observed in the donor, suggests that some PS-inducing factors accumulate in the brain and/or the CSF during PS deprivation.

However, the chemical nature of these factors remains to be investigated and further biological tests have to be performed. Nevertheless, similarly to what was shown recently in the case of the serotonergic system, our working hypothesis proposes that the NA system is involved in the synthesis and/or the liberation of PS-inducing factors. Schematically, these factors would be stored during waking and/or SWS, and "consumed" during PS.

REFERENCES

1 Adrien, J. and Dugovic, C. (1984). *Eur. J. Pharmacol.* **100**, 223–226.
2 Borbély, A.A. and Tobler, I. (1980). *Trends Pharmacol. Sci.* **1**, 356–358.
3 Dement, W.C. (1960). *Science* **131**, 1705–1707.
4 Drucker-Colin, R.R. and Rojas-Ramirez, J.A. (1970). *Brain Res.* **23**, 269–273.
5 Fencl, V., Koski, G., and Pappenheimer, J.R. (1971). *J. Physiol.* **216**, 565–589.
6 Gaillard, J.M. (1983). *Br. J. Clin. Pharmacol.* **16**, 205s–230s.
7 Jacobs, B.L. and Jones, B.E. (1978). In *Cholinergic-Monoaminergic Interactions in the Brain*, ed. Butcher, L.L., pp. 271–290. New York, London: Academic Press.
8 Jouvet, M. (1983). In *Sleep 1982*, ed. Koella, W.P., pp. 2–18. Basel: Karger.

9 Lanfumey, L. and Adrien, J. (1981). *C. R. Acad. Sci.* **292**, 645–647.

10 Lanfumey, L. and Adrien, J. (1982). *Eur. J. Pharmacol.* **79**, 257–264.

11 Lanfumey, L., Dugovic, C., and Adrien, J. (1985). *Electroenceph. Clin. Neurophysiol.* **60**, 558–567.

12 Monnier, M. and Hösli, L. (1964). *Science* **146**, 796–798.

13 Monnier, M., Dudler, L., Gaechter, R., Maier, P.F., Tobler, I., and Schoenenberger, G.A. (1977). *Experientia* **33**, 548–552.

14 Morden, B., Mitchell, G., and Dement, W. (1967). *Brain Res.* **5**, 339–349.

15 Nagasaki, H., Iriki, M., Inoué, S., and Uchizono, K. (1974). *Proc. Japan Acad.* **50**, 241–246.

16 Pappenheimer, J.R., Koski, G., Fencl, V., Karnovsky, M.L., and Krueger, J. (1975). *J. Neurophysiol.* **38**, 1299–1311.

17 Pappenheimer, J.R., Miller, T.B., and Goodrich, C.A. (1967). *Proc. Natl. Acad. Sci. U.S.* **58**, 513–517.

18 Piéron, H. (1913). *Le Problème Physiologique du Sommeil.* Paris: Masson.

19 Pujol, J.F., Mouret, J., Jouvet, M., and Glowinski, J. (1969). *Science* **159**, 112–114.

20 Riou, F., Cespuglio, R., and Jouvet, M. (1982). *Neuropeptides* **2**, 265–277.

21 Sallanon, M., Buda, C., Janin, M., and Jouvet, M. (1982). *Brain Res.* **251**, 137, 147.

22 Schildkraut, J.J. and Hartmann, E. (1972). *Psychopharmacology* **27**, 17–27.

23 Stern, W.C., Miller, F.P., Cox, R.H., and Maickel, R.P. (1971). *Psychopharmacologia* **22**, 50–55.

24 Vimont-Vicary, P., Jouvet-Mounier, D., and Delorme, F. (1966). *Electroenceph. Clin. Neurophysiol.* **10**, 439–449.

20

ORGANIC BROMINE SUBSTANCE: AN ENDOGENOUS REM SLEEP MODULATOR?

NOBUYUKI OKUDAIRA,[*1] HIROAKI KOIKE,[*1] SHIKIO INUBUSHI,[*1] AKIHISA ABE,[*2] MOTOTERU YAMANE,[*2] ISAMU YANAGISAWA,[*2] AND SHIZUO TORII[*1]

*Department of Physiology, Toho University School of Medicine,[*1] and Department of Biochemistry, Tokyo Medical College[*2], Tokyo 143, Japan*

After the discovery of the periodicity of rapid eye movements during sleep (*1*), sleep can be looked upon as a complex and multivariable state. Basically, there are two different types of sleep: non-REM (NREM) sleep and REM sleep. These two sleep states must therefore be separately considered when studying the possible role of humoral factors in the regulation of the sleep-waking cycle. Numerous studies have established that a polypeptide is involved in the promotion of the slow wave sleep (SWS) (*9, 12*). On the other hand, little is known about the REM promoting and/or modulating factor, except for indications for the involvement of a short-chain fatty acid (*6, 14*).

During the course of investigations on bromine intoxication, it was found that, in addition to exogenous bromine, a bromine containing substance existed physiologically both in the cerebrospinal fluid (CSF) and blood (*18*). An organic bromine compound, accounting for the greatest portion of CSF bromine, was isolated by means of silica gel column chromatography from a large amount of CSF pooled in ethanol, and was identified as 1-methyl heptyl-γ-bromoacetoacetate (MHBAA) (*19*).

Subsequently, the bromine level in the CSF during the sleep-waking cycle in encéphale isolé cats was measured using a radioactive method in the CSF. It was found that there were marked, episodic rises in MHBAA level accompanying the onset of REM sleep, while there was no such episodic increase during NREM sleep (*13*).

Recently, we developed a quantitative microdetermination of MHBAA by the radiodilution method combined with fluorometry (*16*). The correlation between MHBAA and REM sleep was reexamined in the cat using the newly developed method. Furthermore, we performed correlative measurements in human plasma.

To see whether the level of intrinsic MHBAA in human blood correlates with REM sleep, we performed the following three studies. The first and second experiments were carried out in man (total sleep deprivation (TSD) and REM sleep deprivation (RSD)), and the third was carried out in the cat.

I. MEASUREMENT OF MHBAA IN HUMAN PLASMA

The amount of MHBAA in plasma was measured by quantitative microdetermination (*16*) as shown in Fig. 1. After mixing 0.5 or 1 ml plasma with a known amount of synthetic 1-methyl heptyl-γ-bromo-(3-^{14}C)-acetoacetate (^{14}C-MHBAA) in methanol, the mixture was reacted with DANSYL-L-cysteine. The DANSYL-L-cysteine derivative of MHBAA was separated by thin layer chromatography (TLC) after treatment of unreacted excess DANSYL-L-cysteine. The ^{14}C-specific activity of the derivative of radioactive MHBAA was measured by the isotopical and fluorometrical technique. The amount of MHBAA in the sample was calculated as follows:

$$\text{The amount of MHBAA (pmol)} = \left(\frac{^{14}\text{C-specific activity of synthetic }^{14}\text{C-MHBAA}}{\text{diluted }^{14}\text{C-specific activity by adding plasma}}\right) - 1$$
$$\times \text{ the amount of the synthetic }^{14}\text{C-MHBAA (pmol)}$$

II. EXPERIMENTAL DESIGN IN HUMAN AND CAT STUDY

The first experiment consisted of a 2-day study. On the first day of the

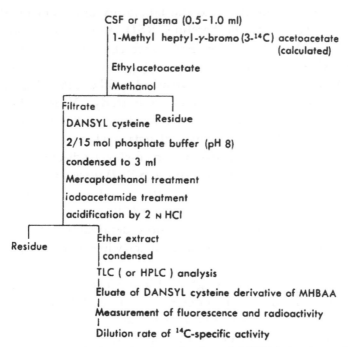

CSF or plasma (0.5–1.0 ml)
1-Methyl heptyl-γ-bromo (3-¹⁴C) acetoacetate
(calculated)
Ethyl acetoacetate
Methanol

Filtrate
DANSYL cysteine   Residue
2/15 mol phosphate buffer (pH 8)
condensed to 3 ml
Mercaptoethanol treatment
iodoacetamide treatment
acidification by 2 N HCl

Residue
Ether extract
condensed
TLC (or HPLC) analysis
Eluate of DANSYL cysteine derivative of MHBAA
Measurement of fluorescence and radioactivity
Dilution rate of ¹⁴C-specific activity

Fig. 1. Quantitative microdetermination of MHBAA by the radiodilution method combined with fluorometry by Yamane and Yanagisawa (*16*).

study, subjects were allowed their usual amount of sleep. One week later, they engaged in the TSD session. Seven healthy male students, ranging in age from 20 to 22 years, volunteered for the experiment. We selected subjects with bedtime between 23:00–0:00 who regularly slept approximately 7 to 8 hr per night. They submitted to a physical examination before the experiment. Each session started at 10:00 and ended at 10:00 on the following day. Subjects were asked to refrain from ingesting any kind of alcohol or other drugs during the two sessions. Blood samples were drawn at 10:00, 16:00, 22:00, and 4:00, after measurements of body temperature, and heart rate, and completing a mood check list. In the first session, the subjects went to bed at 23:00 and were awakened at 7:00. At 4:00 in the morning, blood was taken with minimal disturbance of sleep. In the last session they were totally sleep-deprived throughout the night. In this experiment we did not record a polysomnogram (PSG).

The second experiment was conducted during 4 consecutive nights.

The first night was for adaptation to the laboratory environment, the second, third and fourth for the baseline, RSD, and recovery night, respectively. Throughout the study, PSG were started at 23:00 and ended at 7:00. Nine healthy male volunteers (age range 20 to 28 years, mean 22 years) were recruited. They were asked to fill out the Maudsley Personality Inventory (MPI) test before the study. Venous blood samples were drawn twice, at 22:00 and immediately after the forced wake-up around 7:00. Physical and mental tests, such as recording of body temperature, heart rate, the 31-item forced-choice-adjective-mood-check-list and State-Trait-Anxiety Inventory (STAI) were administered before blood sampling in the night, and after blood sampling in the morning. A further healthy male volunteer, aged 30 years, was recruited for 2 nights of PSG recording without any sleep deprivation. His blood was taken through an indwelling venous catheter while monitoring the sleep stages.

Six male and 22 female cats, weighing 1.9 to 5.5 kg, were sacrificed in the third experiment for the microdetermination of MHBAA in CSF and plasma. Before the collection of CSF and blood, all cats were maintained for more than 4 weeks under diurnal lighting conditions (LD 12:12) with lights on from 6:00 to 18:00, and feeding time at 8:00. Half of them were deprived of REM sleep for 3 days under continuous monitoring of electroencephalogram (EEG), neck electromyogram (EMG), electrooculogram (EOG) and electrocardiogram (ECG). Within 10 min after induction of pentobarbital (Nembutal) anesthesia (50 mg/kg, i.p.), 1 ml of CSF was withdrawn from the cisterna magna and 5 ml of blood was collected from the femoral artery. Samples were collected at time 4:00, 10:00, 16:00, and 22:00. The microdetermination for MHBAA in CSF was measured by the same technique as has been described for plasma determinations.

III.   PLASMA MHBAA VARIATION IN HUMAN AND CAT

A variation in MHBAA concentrations was observed within the same subject over day and night. These were indications for a circadian fluctuation, (*e.g.*, the highest mean value at 10:00 and the lowest one at 22:00), although considerable individual variability was present (**Fig.**

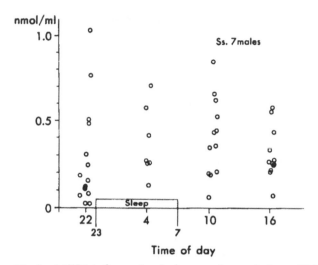

Fig. 2. MHBAA fluctuation in human plasma during a 24-hr period. Seven healthy male volunteers were recruited and blood samples were drawn every 6 hr, at 10:00, 16:00, 22:00, and 4:00 in two sessions. The MHBAA values at 4:00 during TSD session were excluded in this figure. There was a circadian fluctuation in human MHBAA, although considerable individual variability was present.

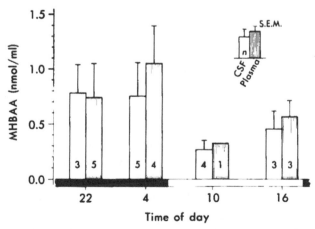

Fig. 3. MHBAA variation in cat CSF and plasma. Note the apparent circadian fluctuation with high levels in the dark period and low levels in the light period.

2). However, there was a clear circadian variation in heart rate and body temperature.

The MHBAA level in cat CSF was similar to that in plasma (0.03–1.88 nmol/ml) and the two parameters showed a significant positive correlation ($r=0.899$, $n=11$, $p<0.005$). Moreover, there was a circadian fluctuation in both MHBAA levels, with the highest value at 22:00 and the lowest at 10:00 (Fig. 3).

According to our microdetermination of MHBAA by the isotope dilution technique, MHBAA showed a circadian fluctuation in both human and cat. In general, there is a clear circadian rhythmicity of the vasopressin and melatonin level in cat CSF, although no obvious circadian rhythm of behavior or sleep was found (7, 10, 11), except for the one report (5). According to our own continuous 48 hr PSG of the cat under LD 12:12 conditions, waking during the dark period (18:00–6:00) was somewhat longer than during the light period (6:00–18:00). This suggested that there might be a circadian rhythmicity in the cat's behavior. Since the highest MHBAA value was obtained at 10:00 in man and at 22:00 in cat, intrinsic MHBAA level appeared to increase during the first half of the wake period. Moreover, MHBAA gradually decreased during the second half of the wake period.

IV. MHBAA LEVEL AND MENTAL STATE

There were some correlations among the mental score, REM sleep and MHBAA level. MHBAA level before and after sleep correlated negatively with the anxiety score in STAI (*i.e.*, the higher MHBAA, the lower anxiety (Fig. 4)). The MHBAA level of the subjects who showed a lower anxiety score in STAI and the extraversion type in MPI was above 0.3 nmol/ml. On the other hand, the level of the others showing a higher anxiety scale and a neurotic and/or intraversion type was below 0.3 nmol/ml (Fig. 4). Moreover, there was a positive correlation between the percent of REM sleep (%SREM) and the anxiety score before sleep ($r=0.44$, $n=16$, $p<0.1$), but no correlation was found between the anxiety and the percent of SWS.

According to Hartmann's theory for the function of sleep (2), a short sleeper who shows a lower anxiety score and the extraversion type, has less REM sleep than the long sleeper, showing a higher anxiety, a

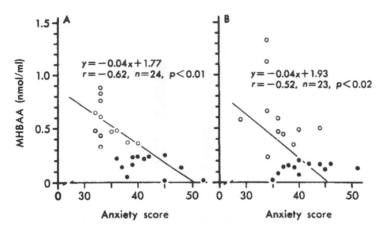

Fig. 4. Correlation between anxiety score and MHBAA level. There was a negative correlation. A: before sleep. B: after sleep. Two groups could be discriminated: A low level group (● below 0.3 nmol/ml) showed a high anxiety. The high level group (○ above 0.3 nmol/ml) showed low anxiety.

neurotic trend, and the intraversion type. Our nine volunteers were neither short nor long sleepers, but variable sleepers. Although they showed a similar tendency in the mental and psychiatry test, there was no consistent change in total sleep time. Moreover, we observed in this study that the high anxiety score before sleep was related to increased REM sleep. After sleep, the low anxiety subjects had a higher MHBAA level and the high anxiety or intraversion type subjects a lower MHBAA concentration.

V. MHBAA LEVEL AND REM SLEEP: EFFECTS OF SLEEP DEPRIVATION

Although the MHBAA level tended to decrease at 4:00 during TSD, it increased at 10:00 following TSD (Fig. 5). We looked for a correlation between MHBAA concentration and %SREM for the time in bed (TIB) during the baseline and recovery night. There was no obvious correlation between %SREM and the MHBAA level in the morning for the 9 subjects. Analyzing the high and low MHBAA groups separately, a significant negative correlation between MHBAA level and %SREM ($r = -0.69$, $n=9$, $p < 0.05$) was obtained in the high MHBAA

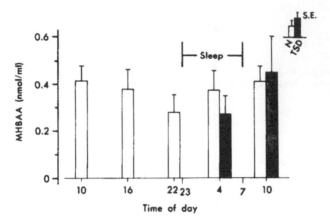

Fig. 5. Effects of TSD on plasma MHBAA level in man. During TSD, MHBAA did not increase, but returned to a high level at 10:00 in the morning.

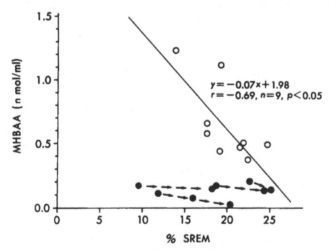

Fig. 6. MHBAA level and REM sleep. There was a negative correlation between percent of REM sleep and MHBAA level, although no correlation existed with the other sleep stages. Dotted arrows represent a decreasing MHBAA level and increasing %REM sleep in the recovery night for each subject.

group. However, no correlation existed with the other sleep stages (Fig. 6). The low MMBAA showed the same tendency within subjects (Fig. 6).

The morning MHBAA level differed between the schedules in the second experiment. The highest value was obtained for the RSD morning, the lowest for the recovery morning (0.48 ± 0.17 and 0.37 ± 0.14 nmol/ml (mean$\pm$S.E.M.), respectively). Surprisingly, in the cat the

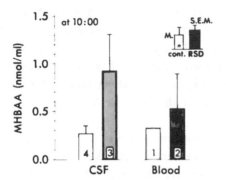

Fig. 7.   MHBAA level in cat CSF and plasma. After 72-hr RSD, the MHBAA level remained high compared to control value.

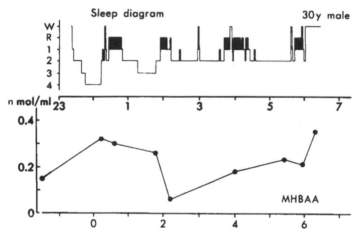

Fig. 8.   MHBAA fluctuation during one night in man. The level tended to increase before REM sleep onset, and to decrease after and/or during REM sleep in each sleep cycle. MHBAA could be expended during REM sleep.

MHBAA level in CSF and blood remained high after the 72-hr RSD, compared to the control value (Fig. 7).

There was a consistent change in MHBAA level before and after REM sleep. The post-REM sleep level was somewhat lower than the pre-REM sleep level (Fig. 8).

In our first experiment, MHBAA showed a minimum level before sleep, and increased gradually toward the morning. On the other hand, MHBAA did not increase during the TSD period. In this connection, we can speculate that MHBAA is sleep-dependent in addition to the circadian rhythmicity (*i.e.*, lower level at sleep onset in men). Since we did not record the PSG in the first experiment, micro-sleep episodes during TSD could not be excluded. All subjects felt sleepy and some of them took a 5- to 10-min nap between 4:00 and 9:00 in the morning. It is reasonable to assume that microsleep or the short nap in the morning during TSD caused the increase in the level of MHBAA at 10:00 (Fig. 5).

Since there was a negative correlation between MHBAA and the percent of REM sleep, we could assume that MHBAA is increased after sleep onset and is expended during REM sleep. After RSD, the MHBAA level was higher than in the baseline and recovery night. In support of the assumption, we observed a correlation between MHBAA and REM sleep in each sleep cycle, even if only in a single subject. MHBAA before REM sleep onset was higher than after REM sleep in each sleep cycle. The MHBAA expense hypothesis during REM sleep was also supported by the study in cat. Thus, we can speculate that MHBAA, being increased after sleep onset, decreases its level during REM sleep. Considering the fact that MHBAA showed a competitive inhibitor of the acetylcholinesterases *in vitro* (*17*), we also speculate that MHBAA modulates REM sleep *via* cholinergic mechanism.

Although we do not have yet precise information concerning the synthetic process of MHBAA, MHBAA could be synthesized outside the central nervous system and transferred into the brain across the blood-brain barrier. Also, we have no data or the direct action of extrinsic MHBAA in animals, except for the acute preparation (*14*). However, we have obtained a REM increasing effect of a MHBAA derivative in young and aged human volunteers (*3, 8*). Thus we can expect that both intrinsic and extrinsic MHBAA can modulate REM sleep considered as a "primitive sleep," (*4*) in animals and man. Further studies concerning the role of authentic MHBAA in the central nervous system and its direct action in chronically prepared animals are necessary.

SUMMARY

One-methyl heptyl-γ-bromoacetoacetate (MHBAA) was measured in both human and cat plasma and cat CSF, using an advanced isotope dilution technique combined with fluorometry. MHBAA, showing a circadian fluctuation in man and cat, significantly correlated with REM sleep. Its level decreased during TSD, but remained high after RSD. In man, MHBAA showed a negative correlation with REM sleep, and with the anxiety score. It is concluded that MHBAA is produced after sleep onset, modulates REM sleep, is expended during REM sleep, and has an anticholinesterase action.

Acknowledgments

We express our thanks to Mr. H. Ohshige, Y. Seki, and K. Kudoh, Lion Corporation Biological Science Laboratory in Odawara, for their help in preparing the cat's study. We also thank Mr. Daniel J. Mullaney for his help in preparing this manuscript.

REFERENCES

1 Aserinsky, E. and Kleitman, N. (1953). *Science* **118**, 273–274.

2 Hartmann, E. (1973). *The Functions of Sleep*. New Haven: Yale Univ. Press.

3 Hayashi, Y., Otomo, E., Okudaira, N., and Endo, S. (1982). *Psychopharmacology* **77**, 367–370.

4 Horne J.A. (1983). In *Sleep Mechanisms and Functions*, ed. Mayes, A., pp. 262–312. Wokingham: Van Nostrand Reinhold UK.

5 Johnson, R.F., Randall, S., and Randall, W. (1983). *J. Interdiscipl. Cycle Res.* **14**, 315–327.

6 Jouvet, M., Cier, A., Mounier, D., and Valtax, J.L. (1961). *C. R. Soc. Biol.* **6**, 1313–1316.

7 Moore-Ede, M.C., Gander, P.H., Eugan, S.M., and Martin, P. (1981). *Sleep Res.* **10**, 298.

8 Okudaira, N., Torii, S., and Endo, S. (1980). *Psychopharmacology* **70**, 117–121.

9 Pappenheimer, J.R., Koski, G., Fencl, V., Karmovsky, M.L., and Krueger, J.M. (1975). *J. Neurophysiol.* **38**, 1299–1311.

10 Reppert, S.M., Coleman, R.J., Heath, H.W., and Kentmann, H.T. (1982). *Am. J. Physiol.* **243**, E489–E498.

11 Reppert, S.M., Schwartz, W.J., Artman, H.G., and Fisher, D.A. (1983). *Brain Res.* **26**, 341–345.

12 Schoenenberger, G.A., Monnier, M., Graf, M., Schneider-Helmert, D., and Tobler, H.J. (1981). In *Sleep-Normal and Deranged Function*, ed. Kamphuisen, H.A.C., Bruyn, G.W., and Visser, P., pp. 25–47. Leiden: Mefar BV.

13 Torii, S., Mitsumori, K., Inubushi, S., and Yanagisawa, I. (1973). *Psychopharmacologia* **29**, 65–75.

14 Torii, S. and Yanagisawa, I. (1980). In *Integrative Control Functions of the Brain*, ed. Ito, M., vol. III, pp. 301–310. Tokyo: Kodansha.

15 Trulson, M.E. and Jacobs, B.L. (1983). *Neurosci. Lett.* **36**, 285–290.

16 Yamane, M. and Yanagisawa, I. (1982). *J. Biochem.* **92**, 2009–2020.

17 Yamane, M., Abe, A., and Yanagisawa, I. (1984). *J. Neurochem.* **42**, 1650–1654.
18 Yanagisawa, I. and Yoshikawa, H. (1968). *Clin. Chim. Acta* **21**, 217–224.
19 Yanagisawa, I. and Yoshikawa, H. (1973). *Biochim. Biophys. Acta* **329**, 283–294.

21

BIOLOGICAL SIGNIFICANCE OF THE CENTRAL NORADRENERGIC SYSTEM: A STUDY OF SKIN CONDUCTANCE RESPONSE AND COCHLEAR EVOKED POTENTIAL

KEN-ICHI YAMAMOTO

Division of Neurophysiology, Psychiatric Research Institute of Tokyo, Tokyo 156, Japan

In the course of a study on schizophrenia, it became inevitable to encounter the problem of the functional significance of the central noradrenaline (NA) system and its relationship to the mechanism of sleep and wakefulness (*10, 13*).

In the periphery, NA is a well-known substance that mediates sympathetic activity. There is considerable evidence that the central NA system is also related to autonomic activity. Electrical stimulation of the locus coeruleus increases the blood pressure and heart rate, and dilates the pupil (*4*). Pharmacological studies have disclosed that central administration of a NA stimulant compound accelerates the cardiovascular activity, whereas administration of a NA depressant decreases its activity (*9*). However, since these organs are innervated reciprocally by the sympathetic and parasympathetic nervous system, these findings do not provide decisive evidence as to whether the central NA system exerts its influence through excitation of the sympathetic nervous activity or through inhibition of the parasympathetic nervous activity. In relation to this problem, the study of the electrodermal response is of advantage, because the sweat gland is an exceptional organ which is innervated only by the sympathetic nerve (7). Our investigation started from an observation on the changes in skin conductance activity after intraventricular administration of 6-hydroxydopamine (6-OHDA).

I. CHANGES IN SKIN CONDUCTANCE ACTIVITY AFTER INTRAVENTRICULAR ADMINISTRATION OF 6-OHDA

Eight adult house cats were trained to wear a one-piece leather suit and to be suspended by the suit from the ceiling of an auditory-insulated experimental box. Skin conductance was measured between the paw pads of both hindlimbs, using silver-silver chloride bioelectrodes (Beckman 650950), sodium chloride paste (Beckman 201210) and adhesive collars (Beckman 650455). The electrodes were connected to a transducer, which had a variable reference resistor and passed a constant voltage (0.5 V) between the two electrodes. The transducer indicated the skin conductance as the ratio to the conductance value of the reference resistor (Fig. 1).

The skin conductance response was evoked by auditory stimuli of 5 kHz frequency 100 dB intensity (C-scale) and one-second duration. They were applied 20 times at irregular intervals ranging from 10 to 60 sec about 25 cm obliquely in front of the animal's forehead. The timing of the stimuli and the skin conductance were recorded on a pen

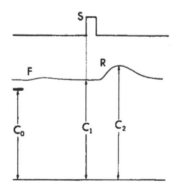

Fig. 1.   Illustration of the skin conductance response and the measurement of its amplitude (change in log conductance). C_0, Conductance value of reference resistor; S, auditory stimulus; R, skin conductance response; F, spontaneous fluctuation. Change in Log. conductance $= \log C_2 - \log C_1 = \log (C_2/C_1)$.

recorder (Nihon-Kohden, WI-680G) and a magnetic tape (Sony Mag-nescale, NFR3915) for later analysis. The skin conductance showed some spontaneous fluctuations (F in Fig. 1). The rate of spontaneous fluctu-ation was passed for the 2-min record preceding the first stimulus and expressed as mean fluctuation number per minute.

After recording the fluctuation and the response to the stimuli, five cats were anesthetized with pentobarbital (26 mg/kg), and 6-OHDA (6 mg dissolved in 300 μl of saline containing 0.1% ascorbic acid) was injected slowly into the lateral ventricle as follows: 1 mg on the first day, followed by 2 mg and 3 mg on the second and third day, respec-tively. The other three cats were treated with the vehicle according to the same schedule. On the first and the seventh day after the final in-jection, the spontaneous fluctuation and the skin conductance response were recorded again. When all the recordings were completed, four 6-OHDA-treated cats and three vehicle-treated cats were anesthetized with pentobarbital and decapitated to estimate the catecholamine con-tent in various regions of the brain by a fluorometric method. Five intact cat brains were used as a control. One 6-OHDA-treated cat was excluded because it died of an illness before the estimation.

The average skin conductance level of the eight intact cats just before the stimulation was 12.4±2.6 (S.E.M.) μmho. When the novel

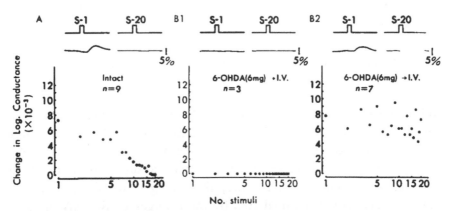

Fig. 2. Specimen recording of the skin conductance response to the first (S-1) and the 20th (S-20) stimulus (top), and the mean amplitude of the response to each stimulus (bottom). A: data from the intact cat. B: data from the 6-OHDA-treated cats one day after treatment. C: data from the 6-OHDA cats seven days after treatment.

auditory stimulus (S in Fig. 1) was given to the cat, the skin conductance increased temporarily after a latency of 1.5 ± 0.1 sec, reaching
its peak within 2.4–5 sec and then declining gradually (R in Fig. 1).
Following repetition of the stimulation, the amplitude of the response
became smaller, eventually disappearing. This phenomenon represents
the "habituation" of the response.

The upper part of Fig. 2A shows an example of the response in an
intact cat at the first (S-1) and the 20th (S-20) stimulus. The lower part
of Fig. 2A indicates the time course of the habituation process. Each
point in this figure represents the mean amplitude of the responses of
eight intact cats (the response amplitude was expressed by "change in
log conductance" from the prestimulus level to its peak value, see Fig.
1). A regression line was computed for each cat on each recording. The
"log habituation point" was determined from the regression line as the
log number of stimuli required for extinction of the response. The average log habituation point of the eight cats was 1.32 ± 0.15.

On the first day after the 6-OHDA treatment, the skin conductance
response was abolished (see Fig. 2B1). The spontaneous fluctuation also
decreased or disappeared.

On the seventh day after treatment, the spontaneous fluctuations
were observed to return to the pretreatment level, but it was noted that
the 6-OHDA-treated cats showed little or no habituation of the skin
conductance response (Fig. 2B2). The vehicle-treated cats showed no
change (*12*).

II. MECHANISM OF THE CHANGES IN THE SKIN CONDUCTANCE ACTIVITY

With respect to the mechanism of the changes in the electrodermal activity, several possibilities should be considered. 6-OHDA is a well-
known neurotoxin that destroys catecholamine nerve terminals selectively and induces compensatory denervation supersensitivity at the
catecholamine synapses. Therefore, the changes in skin conductance
activity may have been mediated by the changes in the NA or dopamine (DA) system, and may be related to the denervation or to the
supersensitivity.

To solve this problem, 80 μg of 6-OHDA was injected bilaterally

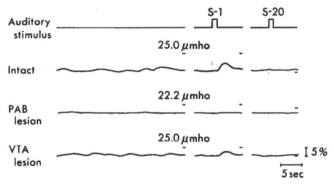

Fig. 3.   The skin conductance activity of an intact, a PAB-lesioned and a VTA-lesioned cat. The uppermost trace indicates the timing of the auditory stimuli.

into the principal adrenergic bundle (PAB) at the part just rostral to the locus coeruleus in five cats. In a further group of six cats, the same dose of 6-OHDA was injected into the ventral tegmental area (VTA). The former injection lowered the cerebral NA content drastically, but did not affect the cerebral DA concentration. The latter treatment reduced the DA content, but produced no change in the NA concentration. Interestingly, the PAB lesion by 6-OHDA obliterated the skin conductance response completely and markedly reduced the spontaneous fluctuation. On the other hand, VTA-lesioned cats showed no change in the skin conductance activity (see Fig. 3) (*11*).

The obliteration of the skin conductance response and the reduction of the spontaneous fluctuation were reproduced by the intraperitoneal injection of propranolol (2 mg/kg) or clonidine (60 μg/kg). Propranolol blocks the NA β-receptor, and clonidine is an agonist of the α_2-autoreceptor of the NA neuron, which is known to reduce the firing rate of the neuron.

The same dose of propranolol (2 mg/kg) restored the habituation of the skin conductance response that had been impaired by intraventricular 6-OHDA. For the obliteration of the skin conductance response of the 6-OHDA casts, 4 mg/kg of propranolol was required.

6-Hydroxydopa is a compound that destroys only the NA nerve terminal and induces supersensitivity of the NA synapse. Intraventricular injection of this drug (1 mg) also resulted in the obliteration of the skin conductance response at 1 or 2 days after the treatment, and in a habituation failure of the response several days after the treatment.

These experimental results suggest that the changes in skin conductance activity are mediated by the NA system rather than the DA system; the obliteration of the skin conductance response and the reduction of the spontaneous fluctuation are probably caused by the denervation of the NA terminals, whereas the habituation failure may be related to the supersensitivity of NA synapses. Thus, elevation of the central NA activity facilitates skin conductance responsivity and the rate of spontaneous fluctuation, while its diminution lowers these activities. The excitation level of the central NA system should be positively correlated to the peripheral sympathetic nervous outflow (13).

III. RELATIONSHIP OF THE CENTRAL NORADRENERGIC SYSTEM TO THE AROUSAL LEVEL

Studies of human electrodermal activity in normal subjects have disclosed that both the habituation rate of the skin conductance response and the spontaneous fluctuation rate reflect the vigilance level of the central nervous system, and that the two parameters are highly correlated (5). If the electrodermal activities are correlated with the central NA activity, a relationship between the central NA system and the vigilance level of the brain should exist. Chu et al. recorded single cell activities of cat locus coeruleus neurons, and analyzed them in relation to the vigilance level of the animal (1). In drowsy wakefulness, the unit fired regularly and slowly at 4.5 ± 1.1 (S.E.M.) per second. Slow wave sleep was associated with a slightly decreased and more irregular neuronal discharge rate of 4.2 ± 1.1 per second. During paradoxical sleep, a bursting discharge was observed in most of the locus coeruleus neurons, the mean firing rate being 10.0 ± 1.3 per second. The same authors reported that the firing rate in attentive wakefulness was even higher (11.6 ± 1.9.).

Electrical stimulation of this nucleus in the unanesthetized and immobilized cat not only elevates the peripheral sympathetic activity, but is also known to desynchronize the cortical electroencephalography (EEG). In our experiment with freely moving cats, chemical stimulation of the NA neuron by yohimbine (0.3 mg/kg) or methamphetamine

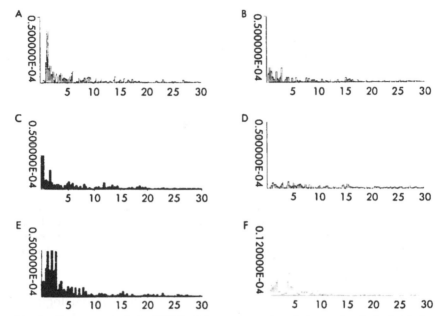

Fig. 4.   Frequency analysis of EEG of the cat cortex in relation to chemical stimulation (C, D) and depression of the NA system. A: EEG pattern during quiet wakefulness. B: EEG pattern while paying attention to a mouse (M(+)). EEG patterns after injection of yohimbine (0.6 mg/kg, M(−)) (C), methamphetamine (0.5 mg/kg, M(−)) (D), clonidine (15.0 μg/kg, M(−)) (E), and propranolol (4.0 mg/kg, M(−)) (F) are compared. Units of the abscissae are Hz and that of the ordinates V^2. Epoch duration of analysis: 4 sec. All data except F are obtained from the same animal.

(0.5 mg/kg) also desynchronized the electrocortical activity (see Fig. 4) (2). Yohimbine is an antagonist of α_2-autoreceptors and consequently increases NA neuronal activity. Methamphetamine is a compound that facilitates the release of catecholamines at the synaptic terminal. By contrast, suppression of the NA neuronal activity by clonidine (15 μg/kg) enhanced the electrocortical synchronization, suggesting increased drowsiness after the administration of this drug (8). With propranolol injection, marked EEG changes could not be noted, even after doses as high as 4 mg/kg.

IV. ROLE OF THE CENTRAL NORADRENERGIC SYSTEM
IN THE CONTROL OF ATTENTION

Propranolol appeared to have some effect on the control of attention.

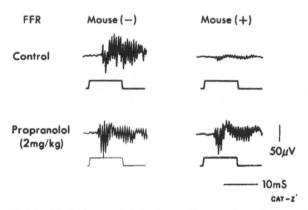

Fig. 5. Evoked potentials in the cochlear nucleus of the cat. The potential is known to reflect the frequency of the auditory stimulus (frequency following response, FFR). FFR is suppressed when the cat pays attention to something other than the auditory stimulus, such as a mouse. This filtering action of attention became weak after the administration of propranolol.

Hernández-Péon was the first to discover the neurophysiological basis of selective attention (3). He demonstrated that click-evoked potentials in the cat's cochlear nucleus became drastically attenuated when the cat was confronted with a mouse. We repeated this experiment by administering pure tones. A pure tone stimulus is known to evoke a frequency-dependent cochlear field potential after a latency of several milliseconds (frequency following response, FFR) (6). Also, in our experiments the FFR was attenuated when the cat was watching a mouse (see Control in Fig. 5). However, this sensory filtering action which was observed when the animal paid attention to some object other than the auditory stimulus, was attenuated after administration of propranolol (see Propranolol in Fig. 5). The abnormal elevation of central NA activity by yohimbine (0.3 mg/kg) and methamphetamine (0.5 mg/kg) increased the distractability of the animal and also impaired the sensory focusing function of attention.

CONCLUSION

Taken together, our experimental findings suggest that the central NA system has an ergotropic function not only on peripheral autonomic activity, but also on the central vigilance level. Thus it can be regarded as an arousal system other than the so-called "diffuse thalamo-cortical activating system." These indications for the dual nature of the arousal regulating system may help us in the analysis of sleep mechanisms.

SUMMARY

Intraventricular administration of 6-OHDA obliterated the skin conductance response in its early stage, and slowed down or completely eliminated the habituation of the skin conductance response in its later stage. Pharmacological analyses have elucidated the mechanism underlying the skin conductance response disappearance and the habituation failure as a denervation supersensitivity and as a compensatory supersensitivity of the NA system, respectively. These results indicate that the NA neuron in the central nervous system is directly related to the sympathetic nervous activity as it is in the peripheral nervous system.

Traditionally, psychophysiologists have assumed that the skin conductance activity reflects the arousal state of the brain and regarded it as an index of arousal level. Chu et al. (1) reported that the firing frequency of the locus coeruleus neuron is strongly correlated to the arousal level during wakefulness. Electrical stimulation of the nucleus locus coeruleus produced EEG desynchronization. Administration of a NA stimulant, such as yohimbine and methamphetamine, also desynchronized the EEG, while clonidine, a NA depressant, enhanced EEG synchronization.

The auditory evoked potential of the cochlear nucleus is inhibited by visual attention which is irrelevant to the auditory stimulus. This inhibition of sensory input by attention was impeded by systemic administration of propranolol, a β-receptor blocker. The sensory filtering function of attention could also be disrupted by NA stimulants, such as methamphetamine and yohimbine.

These experimental findings suggest that the central NA system has an ergotropic function not only on the peripheral autonomic activity, but also on the central consciousness state. It can be regarded as an arousal system other than the so-called "diffuse thalamo-cortical activating system." The assumption of a dual nature of the arousal regulating system may contribute to the understanding of sleep mechanisms.

REFERENCES

1 Chu, N. and Bloom, F.E. (1973). *Science* **179**, 908–910.
2 Florio, V., Bianchi, L., and Longo, V.G. (1975). *Neuropharmacology* **14**, 707–714.
3 Hernández-Péon, R., Scherrer, H., and Jouvet, M. (1956). *Science* **123**, 331–332.
4 Koella, W.P. (1982). *Experientia* **38**, 1426–1437.
5 Lader, M.H. (1980). In *Handbook of Biological Psychiatry*, Part II, ed. van Praag, H.M., Lader, M.H., Rafaelsen, O.J., and Sachar, E.J., pp. 225–248. New York: Dekker.
6 Marsh, J.T., Worden, F.G., and Smith, J.C. (1970). *Science* **169**, 1222–1223.
7 Patton, H.D. (1949). *Proc. Soc. Exp. Biol.* **70**, 412–414.
8 Rotiroti, D., Silvestri, R., de Sarro, G.B., Bagetta, G., and Nistico, G. (1983). *J. Psychiat. Res.* **17**, 231–239.
9 Scriabine, A., Clineschmidt, B.V., and Sweet, C.S. (1976). *Annu. Rev. Pharmacol. Toxicol.* **16**, 113–124.
10 Yamamoto, K. (1982). *Japan. J. Clin. Psychiat.* **11**, 1351–1368 (in Japanese).
11 Yamamoto, K., Arai, H., Moroji, T., and Ishii, T. (1982). *Folia Psychiat. Neurol. Japan.* **36**, 454–455.
12 Yamamoto, K., Hagino, K., Moroji, T., and Ishii, T. (1984). *Experientia* **40**, 344–345.
13 Yamamoto, K. (1985). *Clin. Psychiat.* **27**, 511–520 (in Japanese).

SUBJECT INDEX

AUTHOR INDEX